Toxic Mold Litigation

Edited By
Joel Herz, Esq.
Kimberly Taylor, J.D.

Contributing Authors
Barry L. Bellovin, Esq.
Edward Eshoo, Jr.
Michaela A. Fanning
David M. Governo
Leslie O'Neal-Coble
Stephen Pate, Esq.
Patrick J. Perrone, Esq.
Douglas F. Schleicher, Esq.
Cynthia J. Stephens
Richard J. Valladares, Esq.
Bryan D. Weiss, Esq.
Dennis I. Wilenchik, Esq.

®Lawyers & Judges Publishing Company, Inc.
Tucson, Arizona

Lawyers & Judges Publishing Company, Inc.

P.O. Box 30040 • Tucson, AZ 85751-0040
(800) 209-7109 • FAX (800) 330-8795
e-mail: sales@lawyersandjudges.com

Library of Congress Cataloging-in-Publication Data

Herz, Joel L.
Toxic mold litigation / Joel L. Herz, Kimberly A. Taylor.
p. cm.
Includes bibliographical references and index.
ISBN-13: 978-1-930056-44-2 (pbk.)
ISBN-10: 1-930056-44-3 (pbk.)
1. Personal injuries--United States. 2. Molds (Fungi)--Law and legislation--United States. 3. Actions and defenses--United States. 4. Insurance claims--United States. I. Taylor, Kimberly A. II. Title.
KF3827.R4.H47 2005
346.7303'23--dc22

2005013273

ISBN 1-930056-44-3
ISBN 978-1-930056-44-2
Printed in the United States of America
10 9 8 7 6 5 4 3 2 1

online
www.lawyersandjudges.com

Contents

Acknowledgment

The authors wish to express their appreciation to everyone who played a role in the publication of this book. In particular, we give thanks to the contributors to this publication who withstood our tireless prodding and to the staff at Lawyers and Judges Publishing for their editorial insights. We also must highlight the substantial contribution of Steven Weintraub, Leslie Grant, and Jill Ciotti, whose patience, attention to detail, and management skills have made this book a reality.

Additional thanks must go to our families, friends, and colleagues for providing support and inspiration.

Last, but of course not least, Joel Herz wishes to thank Jane, Jordan, and Abigail for their unending patience, love and support. Also, special thanks goes to his mother, Myra, and his father, Stuart. Kim Taylor wishes to thank Martin most especially for his loving support during the course of this book as well as in life. Thanks also to Sharlene, Ron, Sue, Greg, and all the Taylors for their support in her endeavors, and to Richard for his assistance and support during the early stages of the book.

Joel Herz
Kim Taylor

Introduction

Due to the complicated nature of mold issues, overcoming mold problems means different things to different people. This book was written as both a guide and a resource manual for those facing mold litigation issues. The complexity of addressing mold litigation stems, in part, from the large number of professional disciplines necessary to address mold litigation, as well as the lack of detailed information readily available about newly emerging state laws. Therefore, this book was developed to provide both information and strategic advice to assist parties in mold litigation.

The book is divided into three parts. Part I explains critical background information to put the "mold issue" in context. Part II offers a detailed look at the most important elements of mold litigation in critical States where mold litigation has been prevalent. Finally, Part III addresses insurance issues in mold litigation.

Parts I and III are designed to be read "cover-to-cover" and will provide the reader with not only essential information important to understanding the issues but also with strategic advice toward addressing the problems posed by mold. The authors did not intend for the reader to read Part II cover-to-cover. Rather, this section of the book should serve as critical reference material and offers tremendous insight into any particular state of interest to the reader. Additionally, Part II of the book may be used to compare one state's laws with another state's.

This book is intended to function as a practical reference for persons interested in participating in mold litigation. The authors also intend to update and supplement the book periodically, to keep readers abreast of the latest developments, new laws, and strategies for practically addressing mold issues. ***As this area of law continues to develop at a rapid pace,***

the reader should carefully check the status of any particular law, or regulation described in the book before relying on the accuracy of any particular reference.

Part I

Chapter 1

Mold

Barry L. Bellovin, Esq.

Synopsis

1.1 Introduction

Mold is found virtually everywhere in our environment—both indoors and outdoors and in both new and old structures. Molds are naturally occurring microscopic organisms, which reproduce by spores and have existed since the beginning of time. Any doubt as to the long standing nature of mold problems can be resolved by reference to the Old Testament, Leviticus 14:39-47.

> On the seventh day the priest shall return to inspect the house. If the mildew has spread on the walls, he is to order that the contaminated stones be torn out and thrown into an unclean place outside the town. He must have all the inside walls of the house scrapped and the material that is scraped off dumped into an unclean place outside the town. Then they are to take other stones to replace these and take new clay and plaster the house. If the mildew reappears in the house after the stones have been torn out and the house scrapped and plastered, the priest is to go and examine it and, if the mildew has spread in the house, it is a destructive mildew; the house is unclean. It must be torn down and taken out of the town to an unclean place. Anyone who goes into the house while it is closed up will be unclean until evening. Anyone who sleeps or eats in the house must wash his clothes.

The molds *stachybotrys chartarum* and *aspergilis* have been implicated in some of the most severe injury and property mold claims. Confounding mold litigation is that the effect of mold exposure varies dramatically between individuals. For example, children generally tend to be more susceptible to mold than adults. Those with pre-existing asthma and allergies tend to react more severely than others with no pre-existing condition, all other variables being equal. Confounding the situation further, is that mold exposure can come in combination with exposure to multiple molds and other substances at the same time. In addition, mold exposure varies significantly in terms of the intensity and length in time of exposure. Generally speaking, the lower the duration of exposure and the lower the spore count, the greater chance there is for full recovery of person and property upon removal of the offending mold. Obviously, the reverse tends to be generally true. That is, the longer the exposure and the greater the spore count, the higher chance of illness and damage, assuming a destructive species of mold.

1.2 Legal Aspects of Mold Litigation Overview

When considering an injury or property damage mold claim or a combination thereof, a key consideration is the existence and handling of the evidence. Evidence of mold species and spore count can make or break either a plaintiff's or defendant's case. The best evidence is most often the mold itself, assuming the structure has not been remediated or repaired. The question of how to preserve this evidence should be considered carefully. Chain of custody issues must be considered. Lack of a tightly documented chain of custody places the results in doubt. Any scientific testing results are subject to cross-examination, challenges to legal foundation, and potential exclusion from evidence.

There are different types of industries that have sprung up around the mold issues. The first type of company is a mold testing company, which can consist of an industrial hygienist. These types of company, who are meritorious, carefully gather evidence, document, and use reputable labs to establish to existence of mold, its variety, spore count, and so on. There is another type of company known as a remediation company that resembles a specialized contractor who removes the mold and rebuilds. Mold can be removed in a variety of ways. The classic attack on mold is to stop its source of moisture, remove damaged materials, and dry out the affected environment. There are contractors who can literally seal up a property, and then using a suctioning device can extract the moisture. Remedies dealing effectively with mold continue to be developed.

In an automobile product liability defect case, the key is to secure the automobile. The result of not securing the automobile, often leads to there being no case at all. It is the same with mold litigation. Most often, no product equals no case. The exceptions are few. This is a reality for both plaintiff and defendant. If the plaintiff has evidence of the presence of a harmful species of mold, high spore count, significant exposure time and prompt medical reporting to a qualified physician who can establish causation, they may have a viable case. However, if any of these elements are lacking, the defendant has strong avenues of defense.

A plaintiff who may have suffered mold injury should immediately report their symptoms to their primary care physicians. The plaintiff should also consult the various professionals who are experts in evaluation, diagnosis, and treatment of mold exposure and who are willing to provide testimonial support. The health professionals involved in these types of cases often bear the titles of toxicologist, occupational medicine physicians, and/or pulmonologist. These physicians should provide the necessary diagnostic testing to determine causation. The earlier mold is detected and removed, the better the prognosis for both property and patient. Testing can also rule out the presence of a mold related injury. The types of injuries often claimed to be associated with mold exposure include pulmonary dysfunction, asthma, allergies, chemical pneumonitis, phenomena, headache, rash, mental disorientation and memory loss, and multiple chemical sensitivity. The relationship of these symptoms to mold exposure remains highly controversial. Experts in the medical field differ strongly on whether mold is the culprit for this constellation of symptoms. This leaves ample room for argument by experts on all sides of the table. On the liability side, experts such as contractors, plumbers, landlords, and property managers can be the necessary experts to establish the standards for maintenance, and the breaches thereof. In a defect case, an architect can also be an appropriate expert.

In today's legal climate, experts must be selected with great care. The penalty of disqualification can leave a lawyer with no expert or expert testimony. Ideally, an expert toxicologist, industrial hygienist, or other professional expert will have been previously qualified in a court of law to testify and will have been published in several peer reviewed books, articles or publications. Any expert selected must not only be able to support their opinions, but also the underlying scientific methodology for the assumptions that form the foundations for their opinions. This is a challenge in the mold litigation arena because the science of diagnosis and remediation are still developing. The witness should be able to support the scientific validity of the underlying sampling, data collection, and the efficacy of the equipment used to test and analyze data.

1.3 Legal Theories

Mold cases are different from other types of accidental injury cases in major respects. Mold cases develop slowly, and often silently. Preventing mold often takes due diligence by the property owner. When this due diligence is absent then legal liability for the harm caused can become part of the lawsuit. Mold results from chronic moisture conditions or from the remnants of a flood where the moisture was allowed to remain. It does not take a great deal of moisture to create mold. A small roof leak, leaky plumbing pipe, or cracks or openings around window and door frames are typical sources of water intrusion on a chronic basis leading to mold development. This situation can be compounded by the leaks remaining within walls of the structure. The need for due diligence stems from ongoing maintenance obligations.

There can also be a defect in construction or repair that creates the same obligations and liabilities as lack of maintenance. For example, a defective roof might have a gap in drainage that allows water to escape into the walls. Such a defect can be caused by design error, by roofer error, or both. A maintenance or design defect can easily develop serious consequences by the presence of other hazards. For example, a carpet floor in a bathroom can become dangerously contaminated by a plumbing leak due to the bacterial environment of the bathroom. In addition, a heating and cooling system in a home or office can easily spread contamination throughout the structures and to a larger group of people. The threat of mold can be underestimated in a dry climate. However, dry climates can have significant mold issues due to plumbing leaks which are not climate sensitive, and cracking can occur due to temperature extremes allowing for water intrusion.

There are various legal theories which can be applied through tort and contract theories. With regard to tort theories, negligence is the most prevalent. The negligence typically is for a lack of due diligence due to lack of maintenance. When this lack of maintenance leads to chronic water intrusion and high moisture conditions, which leads to causative injury to person or property, a negligence claim is appropriate. If there are any statues that are violated, negligence per se can be pleaded. Strict liability can also be plead in certain circumstances. Under certain conditions, a nuisance claim may be similarly plead. A nuisance is an interference with the use and enjoyment of property.

Breach of contract and breach of warranty also can and should be plead in a mold claim, where appropriate. If the home had mold that the seller knew about and did not disclose prior to purchase, the buyer has a potential misrepresentation claim. The essences of these claims is that the purchaser did not receive the wind and water tight home and healthful environment that they expected they purchased. General contractors and other building professionals are similarly exposed for defective construction that promotes mold growth. Like a termite infestation that is cured, there is an issue of whether an owner must disclose a remediated mold infestation. Tenants who are subjected to mold may have rights for negligence and breach of contract under their leases. In the context of contractual claims, consideration should also be given to arbitration clauses that might be present. Tenants may also be able to make claims for personal property damaged by mold. Various landlord tenant acts may apply in your jurisdiction. Workman compensation claims, sometimes with third-party liability, can often arise in an office setting where there is the presence of mold.

1.4 Insurance Issues

The applicability of insurance coverage, or lack thereof, is often critical with regard to mold claims and litigation.

In third party claims the insurance industry, in commercial general liability policies, has relied on an absolute pollution exclusion. These clauses generally state that they are not responsible for "bodily injury or property damage arising out of the actual, alleged, or threatened discharge, disbursal, or escape of pollutants at or from premises you own, rent, or occupy." These clauses generally broadly define pollutants. In third party claims the application of this absolute pollution exclusion has been met with inconsistent results. The validity of the exclusion has often depended upon the exact wording contained in the exclusion, the specific policy and the exact facts of the underlying claim.

With regard to first party claims under a homeowner's policy for property damage caused by mold contamination, there is typically a finding of coverage despite pollution exclusion language in the policy. This is especially the case if the mold was the result of a covered loss caused by broken plumbing pipes, or rain damage associated with roof and window leaks.

With the rise in number and magnitude of mold claims, insurers have now sent their policyholders notices of mold exclusions for coverage. Insurers, in an effort to limit their liability, have been especially aggressive in excluding claims for mold from coverage. This presents many challenges for the legal practitioner. The policy language must be researched thoroughly. In addition, one should look to other polices, including umbrella coverage, that does not contain such exclusions. Another source of mold-related litigation is insurance carriers' wrongful refusal to pay claims or their failure to pay the claims fully. In addition, the insurers can also be held liable for a failure to completely and adequately remediate. Contractors who do not properly repair and remediate, can also have independent liability.

1.5 Statute of Limitations

The statute of limitations, is of course different from jurisdiction to jurisdiction. However, one interesting aspect of mold claims is that it can be argued that the statute does not began to run until discovery of the damage or illness.

1.6 Summary

What follows in this book is an in-depth discussion of mold law in various states. The current legal cases and trends are cited and analyzed. The requirements for the admission of expert testimony under current law is also discussed.

Part II

Chapter 2

Evidentiary Standards in Toxic Mold Litigation

Dennis I. Wilenchik, Esq.

Synopsis

2.1 Introduction

It is probably safe to say that most dinner party conversations do not focus on the complexities of mold litigation. Nevertheless, mold is an exciting and evolving subject that has changed the course of litigation. In the last decade, scientists have spent an enormous amount of time and energy in an effort to explain the potential health effects of mold. This research has made most of America think twice about their lives and homes.

Mold exists both indoors and outdoors and typically grows in moist, dark places. There are more than 1,000 different species of mold found indoors, most of which are not harmful. Over the last few years, there has been a lot of talk about toxic mold, including a particular species called *Stachybotrys chartarum*. Like many other molds, *Stachybotrys chartarum* is typically found in water-damaged structures. The media has repeatedly brought to our attention the dangers of mold. Large jury verdicts have also made the headlines and have further created a panic regarding mold. While it is generally thought that mold should be remediated from residences, schools, office buildings, and other structures, recent scientific reports have found no conclusive link between *Stachybotrys chartarum* and injury or disease. These recent studies have greatly affected strategy among parties litigating mold cases because they dictate what kind of information the jury hears. This chapter will focus on the evidentiary standards applicable in mold litigation.

2.2 Expert Witness Evidentiary Standards

If a mold case ever sees a courtroom, both sides will undoubtedly fight to have their expert witnesses testify in front of the jury. At the same time, both parties will attempt to exclude the other side's expert witnesses for obvious reasons. In a mold case, an expert witness may very well be the most important witness to take the stand. An expert witness will be able to show a causal link between the presence of mold and injury or illness or testify to an applicable standard of care or damages. More importantly, a good expert will be able to explain the complexities and subtleties of mold in plain English to the jury. If the trial court excludes such an expert, there may be a large hole in counsel's case. Before an expert can take the stand, however, the trial court judge may act as a "gatekeeper" and exclude the witness. Trial court judges typically have the power to exclude experts under theories which come from two cases: *Frye* and *Daubert*. Most states follow either *Frye* or *Daubert*, but some states follow their own line of cases or statutes.

A. The Frye "general acceptance" test

Frye v. United States[1] was the first major case to establish a standard for the admissibility of new or novel scientific evidence. In fact, *Frye* came years before the enactment of Rule 702 of the Federal Rules of Evidence in 1975. The sole issue on appeal in *Frye* was whether the defense in a murder case could offer an expert witness to testify regarding the systolic blood pressure deception test.[2] The defense proposed to have the expert witness conduct this test and perform it in the presence of the jury, but the Court held that the systolic blood pressure deception test had not gained "general acceptance" in the scientific community to justify admission of expert witness testimony.[3] In other words, the Court did not know whether the test was accurate, and therefore if evidence from it was presented to the jury, it could be reversible error. As part of the holding in *Frye*, the court created the "general acceptance test" and stated that:

> [j]ust when a scientific principle or discovery crosses the line between the experimental and demonstrable stages is difficult to define. Somewhere in this twilight zone the evidential force of the principle must be recognized, and while courts will go a long way in admitting expert testimony deduced from a well-recognized scientific principle of discovery, the thing from which the deduction is made must be sufficiently established to have gained general acceptance in the particular field in which it belongs.[4]

The *Frye* general acceptance standard was prevalent for many years in both state and federal courts up until Federal Rules of Evidence Rule 702 was enacted. Following the enactment of Rule 702 of the Federal Rules of Evidence, some federal district and appellate courts rejected the *Frye* test. The Supreme Court resolved the conflict among these courts in *Daubert v. Merrell Dow Pharmaceuticals, Inc.*,[5] discussed below.

Although a slight majority of states follow the *Daubert* ruling, the 1923 *Frye* decision still remains the law in many state courts. The states that still use *Frye's* "general acceptance" test include Arizona, California, the District of Columbia, Florida, Illinois, Kansas, Maryland, Michigan, Minnesota, Missouri, New York, North Dakota, Pennsylvania, and Washington.[6]

1. The *Frye* procedure

The procedure courts use in deciding whether to allow an expert to testify is rather straightforward. Courts will hold a *Frye* hearing before or during a trial in order to determine whether an expert should testify. The purpose of a *Frye* hearing is to resolve the issue of "general acceptance" of the theories relied on before trial. The advantage of conducting a *Frye* hearing before trial is that the results to the hearing may lead to further settlement negotiations, and the hearing itself takes place out of the presence of the jury. Parties preparing for a *Frye* hearing should ideally prepare notebooks with various scientific reports and case law rulings on similar evidentiary matters.

2. The trial court's discretion under *Frye*

In general, the trial court is afforded great discretion in evidentiary matters and its assessment of the credibility and weight of the expert testimony. An appellate court will review *de novo* a lower court's determination that a scientific principle meets the *Frye* requirement of general acceptance.[7] If the trial court allows an expert to testify at trial, the jury will still be able to decide the appropriate weight to be given to the expert's testimony.

B. The *Daubert* test

In *Daubert v. Merrill Dow Pharmaceuticals, Inc.*, the United States Supreme Court defined a new standard for the admissibility of novel scientific evidence. In this action, the parents of two boys alleged that the prescription anti-nausea drug Bendectin taken during the mother's pregnancy caused birth defects.[8] The evidentiary issue decided by the Court involved whether expert evidence relying on a particular type of blood test was admissible.[9]

At the trial court level, the defendant's expert witness executed an affidavit in which he stated that, based on his review of the published literature, Bendectin did not pose a risk of birth defects.[10] The plaintiffs sought to refute defendant's expert witness' affidavit by introducing the testimony of eight experts who had come to the opposite conclusion.[11] In ruling the plaintiffs' expert testimony inadmissible, the trial judge determined that the plaintiffs' experts' methodologies did not meet the *Frye* general acceptance test.[12] The trial judge granted defendant's motion for summary judgment and the D.C. Circuit Court of Appeals affirmed.[13]

The United States Supreme Court reversed, holding that Rule 702 from the Federal Rules of Evidence superseded *Frye*, and that the general acceptance test set forth by *Frye* was no longer a precondition to the admissibility of scientific evidence.[14] The Supreme Court set forth four criteria for the evaluation of experts: (1) whether the theory or technique has been or can be tested; (2) whether it has been subjected to peer review and publication; (3) the known or potential rate of error; and (4) whether the theory or technique has been generally accepted by the scientific community.[15] These four factors are non-exhaustive but reflect the goal of the Supreme Court to only allow expert witnesses to testify if they are qualified and have a reasonable foundation for their testimony.[16]

The Supreme Court in *Daubert* also created a two-prong test relying on Rule 104(a) of the Federal Rules of Evidence. Trial courts using *Daubert* must (1) make a preliminary assessment of whether the reasoning or methodology underlying the expert's testimony is scientifically valid and applicable to the facts of the case; and (2) decide whether the expert's testimony would be helpful to the jury.[17]

The introduction of *Daubert's* four criteria has prompted a looser standard in admitting scientific evidence, and some critics fear that "junk science" is being admitted into the courtroom. The Supreme Court tried to address these fears by stating its belief that cross-examination, care in instruction on the burden of proof, and presentation of contrary evidence will act as balancing tools against the evidence.[18]

The states that have adopted *Daubert* include Alaska, Arkansas, Colorado, Connecticut, Delaware, Idaho, Indiana, Iowa, Kentucky, Louisiana, Maine, Mississippi, Montana, Nebraska, New Mexico, North Carolina, Ohio, Oklahoma, Oregon, Rhode Island, South Carolina, South Dakota, Tennessee, Texas, Vermont, West Virginia, and Wyoming.[19]

1. Subsequent case law interpreting *Daubert*

The Supreme Court has reaffirmed the *Daubert* decision in subsequent cases. The Court has stated that "abuse of discretion" is the proper standard of review for appeals of *Daubert* rulings,[20] and that experts cannot draw unreasonable conclusions based on well-accepted and reliable methodology. [21] More recently, the Court decided that the ruling in *Dau-*

bert extended to non-technical experts.[22] In December, 2000, an amended version of Rule 702 of the Federal Rules of Evidence went into effect incorporating the *Daubert* decision. Rule 702 of the Federal Rules of Evidence now provides:

> If scientific, technical or other specialized knowledge will assist the trier of fact to understand the evidence or to determine a fact in issue, a witness qualified as an expert by knowledge, skill, experience, training, or education, may testify thereto in the form of an opinion or otherwise, if (1) the testimony is based on sufficient facts or data, (2) the testimony is the product of reliable principles and methods, and (3) the witness has applied the principles and methods reliably to the facts of the case.[23]

State courts, including *Frye* states, may adopt variations of this rule. While the actual rule may vary from state to state, the general premise remains the same: expert witnesses may testify if their specialized knowledge will be helpful to the trier of fact.

2. The *Daubert* procedure

To challenge the admissibility of scientific evidence in a *Daubert* state, an attorney should draft a concise motion pursuant to Rule 702 of the Federal Rules of Evidence. In order to preserve judicial economy, the motion should also include or incorporate all other bases for challenging the admissibility of the evidence. The purpose of a *Daubert* motion is to establish that the proffered evidence lacks scientific validity and reliability. Such a motion often attacks the means of an expert rather than the opinions or conclusions of the expert.

The accompanying memorandum to the motion should provide the court with the general principles of admissibility of expert testimony and discuss each of the four factors enumerated by *Daubert*. If possible, the expert should submit an affidavit in support of the *Daubert* motion. The affidavit should serve to provide the court with admissible, evidentiary bases for excluding the testimony of the expert. After providing the court with a general road map of the case, the affidavit should demonstrate specifically step by step why or how the proffered expert's testimony lacks the scientific validity and reliability required by *Daubert's* four factors.

The Federal Rules of Evidence provide that full hearings outside the presence of the jury are only mandated "when the interests of justice so require."[24] Thus, a preliminary hearing pursuant to Rules 104(a) and (c) of the Federal Rules of Evidence is the most practical approach. If such a hearing is scheduled before trial begins, the result of the hearing could affect settlement negotiations. Some courts are reluctant to spend time conducting a *Daubert* hearing. In those instances, the *Daubert* motion and accompanying documents carry even more importance.

C. States applying a hybrid system or their own evidentiary standards

As stated above, states generally either follow the *Daubert* or *Frye* line of cases. In total, twenty-seven states follow *Daubert* and thirteen states plus the District of Columbia follow *Frye*. This means that ten states either use a combination of both *Frye* and *Daubert* or have devised their own system for determining the admissibility of novel scientific evidence. The states that fall under this category include Alabama, Georgia, Hawaii, Massachusetts, Nevada, New Hampshire, New Jersey, Utah, Virginia, and Wisconsin.[25]

Often times, a particular state's standard is loosely based on *Frye* or *Daubert* or a combination of both. For example, Alabama has neither abandoned *Frye's* "general acceptance" test nor has it adopted the *Daubert* standard. Alabama uses *Frye* with respect to expert scientific testimony on subjects other than DNA,[26] but in dealing with DNA evidence, Alabama uses a statute based on *Daubert*.[27] Hawaii determines the admissibility of evidence based on Rules 702 and 703 of its own Rules of Evidence, which seem to give some credence to both *Frye* and *Daubert*.[28] Because each of these states has their own standards, check with the state's rules of evidence.[29]

D. Rule 703 of the Federal Rules of Evidence

Rule 703 of the Federal Rules of Evidence essentially allows a testifying expert to rely on facts or data not otherwise admissible. Thus, even if the judge does not allow scientific evidence regarding the effects of mold to come in directly, the expert testifying at trial can still base his opinion on such data if the data is reasonably relied on by similar experts in the field. The current Rule 703 of the Federal Rules of Evidence provides:

> The facts or data in the particular cause upon which an expert bases an opinion or inference may be those perceived by or made known to the expert at or before the hearing. If of a type reasonably relied upon by experts in the particular field in forming opinions or inferences on the subject, the facts or data need not be admissible in evidence in order for the opinion or inference to be admitted. Facts or data that are otherwise inadmissible shall not be disclosed to the jury by the proponent of the opinion or inference unless the court determines that their probative value in assisting the jury to evaluate the expert's opinion substantially outweighs their prejudicial effect.[30]

Experts are provided wide latitude in offering opinions, and may state opinions not based on first hand knowledge or observation.[31] Rule 703 may vary among states. As with Federal Rules of Evidence Rule 702, it would be wise to check the local rules of practice. Specifically, some jurisdictions following *Frye* hold that opinion evidence relying on scientific principles *must* meet the *Frye* standards.[32]

2.3 Evidentiary Standards in Mold Cases

Claims of mold-related injury are based on a wide range of allegations of injury. With the boom of mold litigation over the past several years, courts have had many opportunities to decide whether to admit or exclude expert testimony on mold. In evaluating whether to admit such testimony, courts often turn to the *Manual of Scientific Evidence*, which provides an outline in dealing with toxicology subjects like mold.

A. *Manual of Scientific Evidence*

Understanding toxicology, or the science of poisons, is crucial to the science of mold.

> There are three central tenets of toxicology. First, "the dose makes the poison"; this implies that all chemical agents are intrinsically hazardous—whether they cause harm is only a question of dose. Even water, if consumed in large quantities, can be toxic. Second, each chemical agent tends to produce a specific pattern of biological effects that can be used

> to establish disease causation. Third, the toxic responses in laboratory animals are useful predictors of toxic responses in humans.[33]

While some scientists debate the usefulness of testing toxic responses in animals, it is unethical to purposefully expose humans to chemical agents.[34]

In evaluating mold or other poisons, toxicologists consider the hazards presented to the human population and environment and the degree of risk associated with exposure at a specific dose.[35] Furthermore, toxicologists analyze risks, specifically the risk of contracting a disease at a given dose.[36] In the courtroom, toxicologists offer evidence to support or refute a claim that exposure to chemicals caused a plaintiffs' injuries.[37] Courts consider several factors in evaluating the testimony presented by toxicologists.

First, courts consider the species of animal tested along with the biological links to humans.[38] Overall, humans and other mammals share a similar biological bond.[39] Problems arise, however, when humans complain about nonspecific symptoms like nausea and headaches because it is difficult to test laboratory animals for these symptoms.[40] It is also problematic when the laboratory animals differ from humans. For example, rats do not have gall bladders.[41]

Second, courts consider whether the research indicates that a chemical agent affects a specific organ in a laboratory animal, and whether there is a similar link in humans.[42] Toxicologists have found that certain chemical agents only affect a single organ, and some chemical agents affect multiple organs.[43]

Third, courts consider the chemical structure of the compound and its relationship to toxicity.[44] Toxicologists compare the chemical structures of compounds with a known toxicity with chemical structures of compounds with an unknown toxicity, and estimate toxicity based on the molecular similarities between the two compounds.[45]

Fourth, courts consider whether the compound has been subject to in vitro research and whether such findings are relevant.[46] Often times, toxicologists have used in vitro tests to test new products, and in vitro tests often provide early warning detection for toxicity.[47]

Fifth, courts consider whether the association between exposure to a chemical agent and a disease is biologically plausible.[48]

For each mold injury listed below, courts should consider these factors to determine whether an association exists between exposure to mold and risk of inheritance of a disease. These days, any injury claim will be comprised of one or more types of injuries from allergic, infectious, or toxic reactions to mold.

B. Allergic reactions resulting from exposure to mold

Mold allergies and allergic reactions are common and are well documented. The most common types of allergic responses are immunoglobulin E (IgE) and immunoglobulin G (IgG).[49]

The allergic response IgE, which is the more common of the two types, can lead to allergic asthma or allergic rhinitis.[50] People with this type of mold allergy are generally susceptible to a host of other indoor and outdoor allergens, including animal dander, dust mites, and weed, tree, and grass pollens.[51] Mold is just one environmental agent that is problematic for these people. Approximately 40 percent of the population has the allergic response IgE.[52] In addition, approximately 10 percent of the population has allergic antibodies to common inhalant molds and about 5 percent of the population will develop allergic symptoms from molds during their lifetime.[53]

The second allergic response, IgG, can lead to hypersensitivity pneumonitis (HP).[54] "HP results from exaggeration of the normal IgG immune response against inhaled foreign (fungal or other) proteins."[55] Although most cases of IgG derive from exposure to mold at work, some cases have been linked to pet birds, humidifiers, and heating, ventilating, and air conditioning systems.[56]

In sum, allergic reactions to mold, unlike toxic reactions discussed below, are rather common.[57] Courts have often allowed experts to testify regarding allergic reactions.

For example, in *New Haverford Partnership v. Stroot*,[58] the Supreme Court of Delaware affirmed the trial court's decision to allow the admission of the testimony of a physician who opined that the mold found in the apartment of the plaintiffs increased the severity of the plaintiffs' asthma.[59] In admitting this evidence, the Court followed the standard set forth by *Daubert*.[60]

In *Mondelli v. Kendel Homes Corp.*,[61] the Supreme Court of Nebraska, applying *Frye*, held that the trial court abused its discretion in excluding the plaintiffs' proffered expert testimony. The plaintiffs claimed that mold found in their home caused several asthmatic attacks that required hospitalization.[62] The Court found that the plaintiffs' experts were qualified to testify and their testimony was relevant and helpful to the jury.[63]

Not all cases, however, allow expert testimony regarding allergies or asthma. In *Roche v. Lincoln Property Co.*,[64] a United States District Court held that a physician's testimony that mold in a tenants' apartment caused respiratory and other ailments was inadmissible under *Daubert*.[65] The plaintiffs' expert was a doctor board-certified in internal medicine and pulmonary diseases.[66] The court excluded the plaintiffs' expert for three reasons. First, he failed to follow the established methodology for differential diagnosis.[67] Second, he failed to explain how published reports showing a link between mold and allergic reactions related to the plaintiffs.[68] Third, he relied solely on the time frame between the plaintiffs' exposure to molds and the plaintiffs' symptoms without accounting for flu-like symptoms that happened shortly before the plaintiffs moved into the apartment.[69] The plaintiffs' expert failed to "provide any evidence that a particular mold in the [apartment] had a greater possibility than cigarettes, dust mites, fibers [cat dander] . . . or any number of present environmental allergens, of causing health effects in [plaintiffs]."[70]

C. Infection resulting from exposure to mold

Like allergic reactions to mold, infections related to mold exposure are fairly common. Those with compromised immune systems are more susceptible to infections from exposure to mold than most people. Specifically, people undergoing chemotherapy, uncontrolled diabetics, people with advanced AIDS, and people with lymphoproliferative disorders, including acute leukemia, are at risk for fungal infections that invade deep tissue.[71] Only people with severe immune system problems should be concerned about this exposure to mold, and these cases are easy to spot and diagnose.

In addition, certain infections are common to people with more normal immune systems. These infections, albeit less severe than those mentioned in the previous paragraph, are extremely common.[72] Some of these infections include "infection of the feet (*tinea pedis*), nails (*tinea onychomycosis*), groin (*tinea cruris*), dry body skin (*tinea corporis*), and infection of the oral or vaginal mucosa."[73] Often times, the particular fungus involved can be found growing as an indoor mold or on indoor pets like dogs and cats.[74] Like the typical allergic reaction to mold, the infectious reaction is rather mild.

D. Exposure to toxic mold

As noted in the introduction, much of the press coverage surrounding mold in recent years has centered on *Stachybotrys chartarum*. Although *Stachybotrys chartarum* has been found in homes, schools, and offices, other indoor molds also produce mycotoxins, which are metabolites that some people believe cause various injuries and diseases.[75] More importantly, recent scientific publications have "concluded that indoor airborne levels of microorganisms are only weakly correlated with human disease or building-related symptoms and that a causal relationship has not been established between these complaints and indoor exposures to [*Stachybotrys*] *chartarum*."[76] Previous studies regarding infant lung hemorrhages noted *infra* arrived at the opposite conclusion.

"If mycotoxins are to have human health effects, there must be an actual presence of mycotoxins, a pathway of exposure from source to susceptible person, and absorption of a toxic dose over a sufficiently short period of time."[77] At the present time, however, scientists have not established standards for what constitutes harmful levels of mold exposure.[78] While studies have been conducted on rats, mice, and guinea pigs, the results of these tests are not particularly helpful for a plaintiff in a mold case.[79]

Hypothetically, even if a plaintiff could establish such standards for mold exposure, they would not necessarily be relevant because mold found indoors is made up of both indoor and outdoor mold.[80] In addition, outdoor molds are more prevalent and significant in airway allergic disease than indoor molds.[81] And, if outdoor molds are shown to be the "cause" of a plaintiff's problems, the plaintiff has no case for negligence, breach of warranty, or the like.

Scientists have also not found a method to determine exposure levels by examining a specific quantity of mold.[82] In examining a mold sample, the amount of mycotoxin produced depends on several variables, including "the species and strain of mold, length of time the mold has been growing, the amount of available food and water, temperature, and the amount (and even wavelength) of available light."[83] In addition, different varieties of mold move differently.[84] *Stachybotrys chartarum* is a rather idle form of mold that requires some mechanical disturbance in order for it to become airborne.[85] Some laboratories have recently developed procedures to measure the level of mycotoxins in the air, but the level of success of these laboratories is currently unknown.[86]

Finally, no biomarker exists to determine if a person has been exposed to mold.[87] Although some plaintiffs may claim that a high level of a specific antibody exists in their blood work, a small percentage of people "have levels above the reference value for any one tested material. Consequently, a panel of tests has a high probability of producing a false-positive result."[88] In sum, our knowledge regarding toxic mold, including *Stachybotrys chartarum*, is incomplete. Despite the lack of knowledge regarding *Stachybotrys chartarum*, some scientists have developed a protocol in evaluating patients who have possibly been exposed to this toxic mold.[89]

As the following summary indicates, plaintiffs have alleged a large number of injuries and diseases resulting from exposure to mold. Accompanying each description is a summary of the state of the science regarding the disease or injury, as well as the likelihood of admission of such evidence in court under *Daubert* and *Frye*.

1. Infant lung hemorrhage

In 1994, a pediatric pulmonologist alerted the Centers for Disease Control and Prevention (CDC) to eight babies in Cleveland, Ohio who had developed a rare disorder: all eight of these babies had bleeding lungs. In investigating these cases, the CDC noted that the babies all lived in homes with water damage. Thereafter, the investigation centered on mold, particularly *Stachybotrys chartarum*.

In their initial report, the CDC concluded that mold *could* be responsible for the lung bleeding.[90] By 1995, the CDC began conducting press conferences discussing the possible link.[91] In 1997, the CDC issued its second report on the Cleveland case study.[92] Although the second report was still cautious and indicated that further studies were necessary, the media ferociously began to warn America of the "dangers" of mold.[93] Television stations including ABC and CNN devoted news segments to the 1997 report published by the CDC.[94] In addition, print media including the *Wall Street Journal*, the *New York Times*, and the *Chicago Tribune* published articles about the possible link between mold and bleeding lungs,[95] and people created mold awareness websites and Internet message boards to discuss emerging mold issues.

The sensationalism and publicity surrounding the possible link helped lead the CDC to form two expert panels to further study the epidemic.[96] The results from these studies forever changed everything we know about the relationship between mold and human illness. A 1999 report rejected the link between mold and infant pulmonary hemorrhage, stating that the illnesses of the eight infants were not compatible with illnesses known to have been caused by *Stachybotrys chartarum.*[97] The 1999 report also questioned how no other family members exhibited any illness or symptoms related to the mold.[98] An additional report published in 2000 by the CDC further disaffirmed any definitive link between *Stachybotrys chartarum* and pulmonary hemorrhage.[99] Further studies not commissioned by the CDC doubt any such link because *Stachybotrys chartarum* mold spores are unlikely to become airborne.[100] Furthermore, even if *Stachybotrys chartarum* spores were to become airborne, the spores are too large to reach someone's lower respiratory tract.[101]

Based on current research, the possible link between mold and lung hemorrhage cannot satisfy a *Daubert* analysis.[102] Similarly, under a *Frye* analysis, there has been no general acceptance of a purported link between mold and lung damage.

2. Cancer

Any purported link between mold and cancer is even more unrealistic,[103] and purported claims that exposure to *Stachybotrys chartarum* caused cancer are very rare because of the lack of supporting science. While some studies have found that certain fungi and mycotoxins have caused tumors in monkeys, there are only limited studies regarding the carcenogenicity of inhaled aflaxins such as *Stachybotrys chartarum.*[104] These limited studies involving inhaled aflaxins were all flawed in some aspect or another.[105] As such, courts must exclude expert testimony regarding the purported link between mold and cancer as unreliable under both *Daubert* and *Frye.*[106]

3. Brain damage and other cognitive deficiencies

Some courts have allowed experts to testify regarding toxic encephalopathy. Toxic encephalopathy is an impairment that affects intellect, memory, emotions, and personality.[107] In a case decided by a Delaware superior court, the judge allowed the plaintiff's expert to discuss toxic encephalopathy, but allowed the defendants to cross-examine the plaintiff's expert as to any defects in the diagnostic methods.[108] Other courts, however, have excluded evidence regarding toxic encephalopathy. In *Allison v. Fire Insurance Exchange,*[109] the court relied on *Daubert* and its progeny to exclude the testimony of the plaintiff's expert witnesses.[110] One of the plaintiff's experts could not say whether his techniques were generally accepted in the scientific community.[111] Furthermore, both experts' testimony was based on unreliable foundational data.[112] Other courts previously allowed the admission of expert testimony regarding cognitive deficits based on exposure to atypical molds, but these cases are likely irrelevant because they are outdated.[113]

On the other hand, there is some support for brain damage based on people *eating* moldy foods. Centuries ago, people feared ergotism, an illness caused by eating moldy grain products.[114] Ergot is a fungal disease of rye grass. Ergotism is a form of fungal poisoning caused by ingesting ergot.[115] The symptoms of ergotism include muscle spasms, seizures, and hallucinations.[116] In fact, the drug LSD can be extracted from ergot.[117] In the past, people whose diet depended on grain products including rye were susceptible to ergotism.[118] Some scholars propose that ergotism contributed to the Salem Witch Trials in early America and the Black Death in the late middle ages.[119] As recently as 1975, ergotism was reported in India.[120] Nevertheless, there is no scientific support that links the rare disease of ergotism to the inhalation of airborne mold spores.[121] Therefore, any expert testimony regarding a connection between ergotism and inhalation of airborne toxic mold spores must be excluded under both *Daubert* and *Frye*.

4. Fibromyalgia

Most courts will also likely exclude evidence regarding fibromyalgia. In brief, fibromyalgia is a disorder in which people develop muscle and soft tissue pain, tenderness, and fatigue.[122] People with fibromyalgia also report morning stiffness and trouble sleeping.[123] Doctors may have trouble diagnosing fibromyalgia because its symptoms are indistinguishable from symptoms of other disorders.[124] Like many other illnesses mentioned in this section, scientists have not identified a cause associated with fibromyalgia.[125] As such, some courts have chosen to exclude evidence regarding fibromyalgia.[126]

5. Multiple chemical sensitivity

Some plaintiffs have hired expert witnesses to testify that they have multiple chemical sensitivity. Multiple chemical sensitivity is a condition used to describe a person who has been exposed to a number of chemicals, including mold.[127] Although scientists debate whether multiple chemical sensitivity is a "real disorder," some people have complained of symptoms including headaches, fatigue, nausea, confusion, and dizziness.[128] People who claim they have developed multiple chemical sensitivity do not develop antibodies in response to chemical exposure.[129] Most courts have declined to recognize the diagnosis of multiple chemical sensitivity as a valid condition.[130] At the very least, some courts following *Daubert* have not allowed experts to testify regarding multiple chemical sensitivity.[131]

6. Sick building syndrome

Sick building syndrome, often related to multiple chemical sensitivity, refers to a set of symptoms that are attributed to occupancy in a building. Like multiple chemical sensitivity, some courts have refused to allow experts to testify regarding sick building syndrome. Plaintiffs often claim that their illness is a result of working in a poorly ventilated building. Typical symptoms of sick building syndrome include itchy eyes, dry throat, headache, and lethargy.[132] Courts not allowing experts to testify regarding sick building syndrome cite the lack of scientific knowledge regarding the alleged syndrome.

> [N]o specific causative agent has been identified for the symptoms occurring in patients with the sick building syndrome . . . The lack of agreement by workers in this field over the definition of sick building syndrome and inclusion and exclusion criteria for patients suspected of having this syndrome has hampered efforts to design well-controlled studies. Evidence that this syndrome exists as a separate disease entity is weak. Some have claimed that mass hysteria and other psychological factors are responsible for symptoms.[133]

A Delaware court applying *Daubert* placed sick building syndrome in the same boat with multiple chemical sensitivity and refused to allow testimony regarding either because of problems with proof of causation.[134] While some courts have allowed expert testimony in "sick building cases," it should be noted that these cases were published at a time when scientific evidence regarding exposure to mold was less complete.[135]

7. Chronic fatigue syndrome

Unlike sick building syndrome and multiple chemical sensitivity, chronic fatigue syndrome is a diagnosable disease. Chronic fatigue syndrome symptoms often resemble the symptoms of the flu and include headache, fatigue and muscle weakness, mild fever, and difficulty concentrating.[136] Although doctors can diagnose chronic fatigue syndrome, there is no known cause of the disease.[137] In a mold case, when an expert cannot pinpoint the actual cause of chronic fatigue syndrome, expert testimony suggesting that exposure to mold caused the chronic fatigue syndrome should be excluded under both *Daubert* and *Frye*.[138]

8. Reactive airway dysfunction syndrome

Unlike most mold related illnesses and diseases, some courts have allowed experts to testify regarding reactive airway dysfunction syndrome (RADS). RADS is similar to asthma, but is induced by exposure to a high-level irritant, gas, vapor, or fume.[139] Some scientists believe that dusts can cause RADS.[140] In *Minner*,[141] the court allowed expert testimony regarding RADS because the plaintiffs made a sufficient showing that the air quality in the building *could* have caused RADS.[142] Although the court in *Minner* allowed expert testimony regarding RADS, some courts have not admitted expert testimony regarding RADS.[143] Whether exposure to mold can cause RADS is certainly debatable at this point.

2.4 Causation

Plaintiffs bringing mold claims must demonstrate that exposure to mold caused illness or injury. This can be a difficult proposition because, after all, mold is literally everywhere. In order to overcome this burden, plaintiffs in mold cases must demonstrate both general and specific causation.[144]

A. General causation

An expert testifying to general causation must show that a substance like mold is capable of causing a certain physical effect in the general population.[145] "One can prove general causation either through experiments to show that a substance is capable of causing a particular injury or by attempting to show that exposure to a substance increases the risk of a particular injury."[146] As noted earlier, existing scientific studies have not determined a definitive link between exposure to mold and illnesses such as lung hemorrhages, brain damage, and cancer. Therefore, plaintiffs attempting to show general causation with regard to these injuries will be unsuccessful and need not attempt to show specific causation.

In *Allison v. Fire Insurance Exchange*,[147] noted above, two doctors conducted epidemiological studies to examine whether a link existed between exposure to mold and toxic encephalopathy.[148] In evaluating whether epidemiological studies have a reliable foundation, the factors courts examine include whether the study was properly designed, biased, capable of repetition with the same results 95 percent of the time, and whether exposure to the substance more than doubles the risk of injury.[149] The court in *Allison* determined that the epidemiological tests used by the plaintiffs' experts failed, could not be repeated with the same results 95 percent of the time, and could not show that exposure to the substance more than doubled the risk of injury.[150] The court therefore found that there was no reliable foundation for the general causation evidence and, as a result, did not even address the issue of specific causation.[151]

B. Specific causation

Specific causation requires that the plaintiff prove that the substance actually caused the plaintiff's alleged injury.[152] One of the biggest hurdles in proving specific causation is to show that a plaintiff was exposed to a sufficient level of mold over a sufficient amount of time to cause the injury claimed.[153] Mold exists in virtually every home or building in some quantity. Although almost any substance can be toxic above a certain level of exposure, certain molds like *Stachybotrys chartarum* do not normally become airborne. Moreover, scientists have not developed standards for harmful mold exposure.

Nevertheless, in order to prove specific causation in a mold case, some commentators suggest that the plaintiff (1) properly identify the mold and mold byproduct allegedly causing a plaintiff's injury; (2) determine the plaintiff's proximity to the mold; (3) determine the amount of time the plaintiff was exposed to the mold; (4) determine the alleged exposure pathway; and (5) analyze the plaintiff's medical history including the specific details of the plaintiff's symptoms and the corresponding timeframes.[154]

C. Case law illustrating causation principles

Most reported cases evaluating causation have determined that causation had not been established. For example, in *Roche v. Lincoln Property Co.*,[155] the plaintiffs claimed that mold exposure caused them to develop certain respiratory ailments.[156] The plaintiffs moved into their apartment on April 4, 2001, but one of the plaintiffs had related health problems *prior* to moving into the apartment.[157] In deciding to exclude the plaintiffs' expert witnesses, the court noted that one plaintiff had asthma since she was a child, and visited the hospital for flu-like symptoms several times before moving into the apartment.[158] That particular plaintiff even visited the emergency room complaining of bronchitis on March 4, 2001, less than one month before moving into the apartment.[159]

Similarly, in an unpublished decision from California, an appellate court found that causation did not exist. In that case, the plaintiff claimed that a broken sewer pipe outside her townhome and mold in her garage caused her bronchitis.[160] The plaintiff saved the contents of her vacuum cleaner bag and a filter from her air purifier, and later tested these samples to show causation.[161] The trial court found that the link between the evidence, tested eighteen months after the plaintiff collected it, and the plaintiff's illness was too tenuous.[162] The California Court of Appeals sided with the lower court but remanded the case to determine if the premises became uninhabitable and to calculate the amount of rent to be abated.[163]

An unpublished Ohio opinion also declined to find causation. In that case, the plaintiff alleged that defendants negligently and recklessly had contributed to the damp and moldy conditions in her basement office space, which caused her lung disease.[164] The plaintiff worked for a nonprofit organization that rented office space in a basement from defendants at *no charge* as long as the nonprofit organization assumed responsibility to care for the space.[165] In her absence, the plaintiff claimed that mold growth accumulated in the basement.[166] The plaintiff claimed that the mold in the basement caused her lung disease, but the court found that basements, by nature, are more conducive to mold growth than other areas.[167] The defendants did not act in a willful, wanton, or reckless manner, and the fact that mold grew in a basement did not give rise to a dangerous condition.[168] The trial court granted defendant's motion for summary judgment and the appellate court affirmed.[169]

Some courts have denied motions for summary judgment brought by plaintiffs because a material issue of fact existed as to causation. In *Watters v. Department of Social Services*,[170] the Louisiana Court of Appeals found that a genuine issue of material fact existed as to whether an office-building worker was exposed to mold.[171] Approximately 680 plaintiffs had brought a class action against the Department of Social Services alleging injuries as a result of being exposed to mold and asbestos in an office building.[172] Plaintiff Christine Songy filed a motion for summary judgment and the trial court denied the motion.[173] One of the plaintiff's experts indicated that she was exposed to mold, but he could not determine if the exposure occurred at the office building in question.[174] Additional experts also testified at depositions that the plaintiff was exposed to mold.[175] On review, the Court of Appeals concluded that a genuine issue of material fact existed because there was no definitive finding of mold in any area of the building that the plaintiff frequented.[176] Therefore, the trial court properly denied the plaintiff's motion for summary judgment on the issue of whether she was exposed to mold at work and whether her exposure to mold was a contributing factor to her injuries.[177]

In an unpublished decision from Michigan, in granting defendants' motion to exclude expert testimony, the court found that "there is not enough evidence to sustain the proposition that mold exposure causes specific medical diseases or illnesses."[178] Plaintiffs were eleven individuals living in a manufactured home park owned and operated by defendants who claimed that they suffered property and personal injuries due to mold growing inside their homes resulting from flooding and water drainage problems.[179] "Where the scientific community is conflicted about its ability to establish standard [sic] related to mold exposure, a court cannot be a place for applying and deliberating on such speculative standards."[180] The judge also chastised the media for its role publicizing the mold phenomenon. "[o]nce a claim is made publicly as certitude, it is difficult to reclaim reason and rational resolution."[181]

Not all courts, however, have ruled in favor of defendants on the issue of causation. In *New Haverford Partnership v. Stroot*,[182] the Supreme Court of Delaware held that the expert opinion, of a physician certified in environmental medicine, that excessive mold in apartments *proximately caused* tenants' health problems was admissible.[183] Although the physician neglected to conduct tests to exclude other possible causes of the plaintiffs' health problems, the physician followed the scientifically accepted procedure of obtaining medical histories and patient questionnaires.[184] The court also found that the foundation for an expert's causation opinion need not be established with the precision of a laboratory experiment.[185] The plaintiffs' expert did not conduct extensive baseline tests regarding the mold levels inside and outside the apartment building.[186] The court found the plaintiffs' expert's oversight went to the weight and not the admissibility of the opinion.[187]

Similarly, in *Liska v. Travelers Property Cas.* Corp.,[188] an insurance coverage case, a Massachusetts court found an expert's methods to be sufficiently reliable to be admissible. Plaintiff's expert relied on the fact that mold was present in plaintiff's home, plaintiff's blood tests indicated that she had been exposed to mycotoxins, and plaintiff's symptoms were not present when she moved into the house but subsided after she moved out.[189]

2.5 Preparing Data for Trial

The New York City Department of Mental Health and Hygiene proposed guidelines for inspecting and sampling mold in 1993.[190] These guidelines were developed in response to mold growth problems in several New York City buildings in the 1990s. The report initially proposed that residents visually inspect and assess any problem areas. In particular, residents should inspect ventilation systems, ceiling tiles, wallboard, cardboard, and other paper surfaces for evidence of mold. The report also suggests that trained professionals obtain bulk or surface samples to identify the particular fungal contaminants. According to the report,

> [b]ulk samples are usually collected from visibly moldy surfaces by scraping or cutting materials with a clean tool into a clean plastic bag. Surface samples are usually collected by wiping a measured area with a sterile swab or by stripping the suspect surface with clear tape. Surface sampling is less destructive than bulk sampling. Other sampling methods may also be available. A laboratory specializing in mycology should be consulted for specific sampling and delivery instructions.[191]

With regard to air monitoring samples, the report stated that if a resident chose to monitor air samples inside his home, he should concurrently take an air sample from outside.[192] The report further stated that someone trained in air sampling methods for microbial contaminants should perform the test and that a laboratory specializing in mycology should analyze the air samples.[193] The report, however, did not recommend that everyone conduct air samples of their homes for contamination. Air monitoring is appropriate if the resident has been diagnosed with a disease that may be associated with exposure to mold or fungus, if the resident suspects that the ventilation systems are contaminated, or if the resident suspects mold is present but cannot visually verify the presence of mold.[194]

Once an air sample is collected, the sample should be sent to a laboratory, preferably one accredited by the American Industrial Hygiene Association (AIHA).[195] While laboratories can compare the indoor samples against the outdoor samples, there are no guidelines in the industry. Laboratory test results by themselves are not sufficient to offer any medial diagnoses.[196]

As of this publication, over ten years after these guidelines were proposed, there are still no uniform standards or findings concerning "safe" or "unsafe" levels of mold exposure. Furthermore, the mere presence of mold in a building does not necessarily mean that mycotoxins are present, or that they are present in any harmful quantities.[197] Indeed, the American College of Occupational and Environmental Medicine concluded that inhalation of a toxic dose of mycotoxins is "highly unlikely at best, even for the hypothetically most vulnerable subpopulations."[198]

2.6 Conclusion

The state of the science regarding toxic mold is, at best, inconclusive. While several years ago, scientists claimed a link between *Stachybotrys chartarum* and disease or injury, scientists have since backed away from that conclusion. At the moment, scientists continue to research the potential harms related to toxic mold, but have not published any definitive findings. In order to avoid admitting "junk science" into evidence, courtrooms have to be cautious by not allowing evidence regarding toxic mold in front of the jury. Whether a court follows the *Daubert* or *Frye* methodology, or its own hybrid standard, courts must act as gatekeepers and deny the admission of such evidence, even with their broad discretion.

Regardless of the difficult evidentiary hurdles plaguing plaintiffs, the threat of mold exposure should be taken seriously. People discovering mold in homes, schools, offices, or other buildings should take careful steps to remediate the problem. In addition, professionals with the proper equipment should perform remediation procedures. As long as scientists debate the threat of mold exposure, people should take steps to protect themselves and their families.

Endnotes

1. *Frye v. United States*, 293 F. 1013 (D.C. Cir. 1923).

2. *Id.*

3. *Id.* at 1014.

4. *Id.*

5. *Daubert v. Merrell Dow Pharmaceuticals, Inc.*, 509 U.S. 579 (1993).

6. 90 A.L.R.5th 453 (2001).

7. *See, e.g.*, *State v. Marshall*, 193 Ariz. 547, 549, 975 P.2d 137, 139 (App. 1998).

8. *Daubert*, 509 U.S. at 579.

9. *Id.* at 582.

10. *Id.*

11. *Id.* at 583.

12. *Id.*

13. *Id.* at 583–84.

14. *Id.* at 579.

15. *Id.* at 593–94.

16. *Id.* at 594.

17. *Id.* at 592–93.

18. *Id.* at 596.

19. 90 A.L.R.5th 453 (2001).

20. *General Electric Co. v. Joiner*, 522 U.S. 136, 143 (1997).

21. *Id.* at 146.

22. *Kumho Tire Co. v. Carmichael*, 526 U.S. 137, 147–149 (1999).

23. Fed. R. Evid. 702.

24. Fed. R. Evid. 104(c).

25. 90 A.L.R.5th 453 (2001).

26. *Parker v. State*, 777 So. 2d 937 (Ala.Crim.App. 2000); *Hoosier v. State*, 612 So.2d 1352 (Ala.Crim.App. 1992).

27. ALA. CODE § 36-18-30.

28. *State v. Ito*, 90 Hawaii 225, 978 P.2d 191 (App. 1999) (the following factors should be considered in determining the admissibility of scientific evidence at trial: (1) the evidence will assist the trier of fact to understand the evidence or to determine a fact in issue; (2) the evidence will add to the understanding of the jury; (3) the underlying theory is generally accepted as valid; (4) the procedures used are generally accepted as reliable if properly performed; (5) the procedures were applied and conducted properly in the present instance; and (6) whether admitting such evidence will be more probative than prejudicial).

29. 90 A.L.R.5th 453 (2001).

30. Fed. R. Evid. 703.

31. *Daubert v. Merrell Dow Pharmaceuticals, Inc.*, 509 U.S. 579, 591 (1993). *See also, Gussack Realty Co. v. Xerox Corp.*, 224 F.3d 85, 94-95 (2d Cir. 2000) (Even though experts did not conduct their own tests and relied primarily on data provided others, testimony of expert admissible under *Daubert* standard).

32. *See State v. Boles*, 183 Ariz. 563, 905 P.2d 572 (App. 1995) (when *Frye* test is applicable, opinion testimony not based on generally accepted principles is inadmissible regardless of the fact that the expert rendering opinion is subject to cross-examination).

33. Bernard D. Goldstein and Mary Sue Henifin, *Reference Guide on Toxicology: Reference Manual on Scientific Evidence*, 2nd ed. 401, 403 (2d ed. 2000) available at http://air.fjc.gov/public/fjcweb.nsf/pages/16.

34. *Id.* at 405.

35. *Id.* at 403.

36. *Id.*

37. *Id.* at 404.

38. *Id.* at 419.

39. *Id.* at 420.

40. *Id.*

41. *Id.*

42. *Id.*

43. *Id.*

44. *Id.* at 421.

45. *Id.*

46. *Id.* at 422.

47. *Id.*

48. *Id.*

49. American College of Occupational and Environmental Medicine (ACOEM), *Adverse Human Health Effects Associated with Molds in the Indoor Environment* (2002) available at http://www.acoem.org/position/statements.asp?CATA_ ID=52.

50. ACOEM, *supra* note 49. Allergic rhinitis includes nasal discharge, sneezing, and conjunctivitis. O. Edwin McClusky, *Black Mold and Human Illness*, Report of Council on Scientific Affairs. Asthma includes bronchospasm, wheezing, mucous secretion, and plugging. *Id.*

51. ACOEM, *supra* note 49.

52. *Id.*

53. *Id.*

54. *Id.* Hypersensitivity pneumonitis includes farmer's lung (a lung tissue inflammation occurring from exposure of an inhaled antigen) and allergic aspergillosis (a rare lung tissue inflammation involving both airways and tissues in the lungs). O. Edwin McClusky, *supra* note 50.

55. ACOEM, *supra* note 49, *citing* J. Fink and M.C. Zacharisen, "Hypersensitivity Pneumonitis." In: E. Middleton Jr. et al. eds., *Allergy: Principles and Practice* (St. Louis: Mosby, 1998) 994–1004.

56. ACOEM, *supra* note 49.

57. *See*, e.g., IOM Committee on Damp Indoor Spaces and Health, Board on Health Promotion and Disease Prevention, *Damp Indoor Spaces and Health* at 194, 201, 208, 226 available at http://www.nap.edu/books/0309091934/html/ (sufficient evidence of an association between presence of mold in a damp indoor. environment and upper respiratory tract symptoms, cough, wheeze, dyspnea (shortness of breath), and asthma (in sensitize asthmatic people).

58. *New Haverford Partnership v. Stroot*, 772 A.2d 792 (Del. 2001).

59. *Id.* at 796, 799.

60. *Id.* at 799.

61. *Mondelli v. Kendel Homes Corp.*, 262 Neb. 263, 631 N.W.2d 846 (2001).

62. *Mondelli*, 262 Neb. at 266, 631 N.W.2d at 852.

63. *Mondelli*, 262 Neb. at 270-276, 631 N.W.2d at 854-859.

64. *Roche v. Lincoln Property Co.*, 278 F.Supp.2d 744 (D.Va. 2003).

65. *Id.* at 750–52.

66. *Id.* at 749.

67. *Id.* at 746.

68. *Id.*

69. *Id.*

70. *Id.* at 765 *citing Flores v. Allstate Texas Lloyd's Co.*, 229 F.Supp.2d 697, 702 (S.D. Tex. 2002).

71. ACOEM, *supra* note 49.

72. *Id.*

73. *Id.*

74. *Id.*

75. *Id.*

76. *Id.*, *citing* R.S. Tobin et al., "Significance of Fungi in Indoor Air: Report of a Working Group," *Can. J. Public Health* 78:S1–S32 (1987); D. Menzies and J. Bourbeau, "Building-Related Illnesses," *N. Engl. J. Med.* 337:1524–31 (1997); F. Fung, R. Clark and S. Williams, "Stachybotrys: A Mycotoxin-Producing Fungus of Increasing Toxicologic Importance," *J. Toxicol. Clin. Toxicol.* 36: 79–86 (1998); C.A. Robbins et al., "Health Effects of Mycotoxins in Indoor Air: A Critical Review," *Appl. Occup. Environ. Hyg.* 15: 773–84 (2000); D.L. Sudakin, "Stachybotrys Chartarum: Current Knowledge of Its Role in Disease," *Med. Gen. Med.* E11 (2000); E.H. Page and D.B. Trout. "The Role of Stachybotrys Mycotoxins in Buildings Related Illness," *Am. Ind. Hyg. Assoc. J.* 62:644–8 (2001); Al Terr, "Stachybotrys: Relevance to Human Disease," *Ann. Allergy Asthma Immunol.* 87:57–63 (2001); H.A. Burge, "Fungi: Toxic Killers or Unavoidable Nuisances?" *Ann. Allergy Asthma Immunol.* 87:52–56 (2001); *see also* IOM Committee, *supra* note_, at 253-254

77. ACOEM, *supra* note 49.

78. Cliff Hutchinson and Robert Powell, *A New Plague—Mold Litigation: How Junk Science and Hysteria Built an Industry*, commissioned by the U.S. Chamber Institute for Legal Reform and the Center for Legal Policy at the Manhattan Institute (2003), page 27.

79. ACOEM, *supra* note 49.

80. Hutchinson, *supra* note 77, at 27.

81. ACOEM, *supra* note 49.

82. Hutchinson, *supra* note 77, at 27.

83. *Id., citing* Jonathan S. Rutchick, "Evaluating Mold Exposure and Neurologic Complaints: No Easy Answers," *Claims* magazine, March, 2003, p. 53.

84. Hutchinson, *supra* note 77, at 27 *citing* Harriet A. Burge, "Fungi: Toxic Killers or Unavoidable Nuisances?" *Ann. Allergy Asthma Immunol.* 87:33, 36 (2001).

85. *Id.*

86. For example, http://www.aerotechlabs.com offers a service to measure the amount of *Stachybotrys chartarum.*

87. Hutchinson, *supra* note 77, at 28.

88. ACOEM, *supra* note 49.

89. The protocol includes the following steps:

1. A thorough history, including the duration and nature of symptoms, home environmental and workplace history, past medical history, and family history
2. A physical exam
3. Exclusion of more common infectious causes
4. Phenotyping the patient as atopic versus non-atopic, which includes skin testing
5. Perform a chest x-ray or high-resolution CT of chest
6. Supportive testing including serologic testing for specific IgG, IgE, or IgA to mold, hypersensitivity pneumonitis screen, and consideration of humoral and cell-mediated immune system evaluation
7. Environmental assessment including walkthrough, air sampling, and measurement of known perennial allergens, irritants, dew point, and mycotoxins
8. Measurement of total symptom scores in and out of the environment
9. Measurement of peak expiratory flow rates in and out of the environment event every two to three hours while awake, and correlation with environmental exposure measurements
10. Consideration of specific provocation test

O. Edwin McClusky, *supra* note 50, *citing* J.A. Bernstein, "The Role of the Allergist in Building Related Illness," in: American Academy of Allergy, Asthma, and Immunology 58th Annual Meeting handout materials (2002).

90. Hutchinson, *supra* note 77, at 12.

91. Hutchinson, *supra* note 77, at 12 *citing* "Science Update," *Dallas Morning News*, April 3, 1995, at 8D; *see also*, *Cleveland Plain Dealer*, September 14, 1995, at 12B (researchers "99.999 percent certain they've pinpointed" fungus as cause).

92. *See* "Update: Pulmonary Hemorrhage/Hemosiderosis among Infants: Cleveland, Ohio 1993–1996," *Morbidity & Mortality Weekly Report* 46(2):33 (1997).

93. Hutchinson, *supra* note 77, at 12.

94. *Id.*

95. *Id.*

96. *Id.*

97. *Id.* at 14, citing *Report of the CDC Working Group on Pulmonary Hemorrhage/Hemosiderosis* (June 17, 1999).

98. Hutchinson, *supra* note 77, at 14.

99. *Id.* citing "Update: Pulmonary Hemorrhage/Hemosiderosis Among Infants: Cleveland, Ohio, 1993–1996," *Morbidity & Mortality Weekly Report* 49(09): 180 (2000).

100. *See e.g.*, *Clearing the Air: Asthma and Indoor Air Exposures* at 163, National Academy of Sciences, Institute of Medicine (2000).

101. Hutchinson, *supra* note 77, at 30, *citing* D.M. Kuhn and M.A Ghannoum, "Indoor Mold, Toxigenic Fungi, and *Stachybotrys Chartarum*: Infectious Disease Perspective" *Clinical Microbiology Reviews* 16(1):144, 159 (2003)

102. Hutchinson, *supra* note 77, at 30; *see also*, IOM Committee, *supra* note_ , at 242-243.

103. *Id.*

104. *Id. citing* Kuhn, *supra* note 100 at 144, 159–162 (2003)

105. Hutchinson, *supra* note 77, at 30.

106. *Id*; *see also*, IOM Committee, *supra* note_, at 250-251.

107. *Minner v. American Mortgage & Guaranty Co.*, 791 A.2d 826, 857 (Del. Super. 2000) *citing* WHO & Nordic Council of Ministers, *Chronic Effects of Organic Solvents on the Central Nervous System and Diagnostic Criteria*, 8 (1985).

108. *Id.*

109. *Allison v. Fire Insurance Exchange*, 98 S.W.3d 227 (Tex. App. 2002).

110. *Id.* at 239–40.

111. *Id.* at 239.

112. *Id.* at 240.

113. *New Haverford Partnership v. Stroot*, 772 A.2d 792 (Del. 2001) (applying *Daubert*, the Supreme Court of Delaware found that evidence opined by experts was accepted in the scientific community).

114. Hutchinson, *supra* note 77, at 28.

115. "Poisons of the Past: Ergotism" at http://www.uic.edu/classes/osci/osci590/5_1Poisons%20of%20the%20Past%20Ergotism.htm.

116. Hutchinson, *supra* note 77, at 28–29 citing Kuhn, *supra* note 103 at 159-60.

117. "Poisons of the Past: Ergotism," *supra* note 114.

118. *Id.*

119. *Id.*

120. Hutchinson, *supra* note 77, at 29, *citing* Kuhn, *supra* note 100 at 159.

121. Hutchinson, *supra* note 77, at 29.

122. *Fibromyalgia Topic Overview*, available at http://my.webmd.com/content/healthwise/10/2429.htm?lastselectedguid={5FE84E90-BC77-4056-A91C-9531713CA348}.

123. *Id.*

124. *Id.*

125. *Id.*

126. *See, e.g.*, *Minner v. American Mortgage & Guaranty Co.*, 791 A.2d 826, 855 (Del. Super. 2000) ("This sort of speculative testimony in stating that the environmental material in the building was a causal factor in the Plaintiffs' development of [fibromyalgia], without the sufficient exclusion of other causative factors, is precisely the type of testimony that should be kept from the jury under the principles of *Daubert*").

127. "Multiple Chemical Sensitivity" at http://my.webmd.com/content/article/61/67473.htm?lastselectedguid={5FE84E90-BC77-4056-A91C-9531713CA348}.

128. *Id.*

129. *Id.*

130. *See, e.g.*, *Minner*, 791 A.2d at 850.

131. *Id.* ("Dr. Ziem's testimony concerning MCS must be excluded from the jury as unreliable on two fronts under the *Daubert* standard. First, the Court is satisfied that MCS is not a scientifically valid diagnosis. Second, the Court opines that Dr. Ziem's testimony that the Discover Card building somehow caused Plaintiffs' MCS seems to be based on nothing other than speculation"). *See also*, *Bradley v. Brown*, 42 F.3d 434 (7th Cir. 1994) (testimony of physician that multiple chemical sensitivity was the result of pesticide exposure properly excluded); *Summers v. Mo. Pac. R.R.*, 132 F.3d 599, 603 (10th Cir. 1997) ("MCS is a controversial diagnosis that has been excluded under *Daubert* as unsupported by sound scientific reasoning or methodology.").

132. "Are You a Victim of Sick Building Syndrome?" Available at http://my.webmd.com/content/article/35/1728_97473.htm?lastselectedguid={5FE84E90-BC77-4056-A91C-9531713CA348}.

133. *Minner*, 791 A.2d at 852 *citing* "Clinical Ecology-Council on Scientific Affairs," *JAMA* 268: 3455, 3466 (1992).

134. *Minner*, 791 A.2d at 867–68.

135. *See, e.g., Centex-Rooney Const. Co., Inc. v. Martin Country*, 706 So.2d 20 (Fla. App. 1998) (Relying on *Frye*, the Florida Appellate Court affirmed the lower court's decision to admit the expert testimony of two doctors who suggested the existence of health hazards based on toxic mold found in several buildings. The Florida Appellate Court's decision to allow two experts to testify was not based on any blanket rule on the admissibility of "toxic mold" testimony but on the fact that the experts in question provided specific references to published medical support of their claims and methodology at a *Frye* hearing.).

136. "Newly Diagnosed: Chronic Fatigue Syndrome (CFS)," available at http://my.webmd.com/content/article/70/81020.htm?lastselected guid={5FE84E90-BC77-4056-A91C-9531713CA348}.

137. *Id.*

138. *See, e.g., Minner*, 791 A.2d at 853-54.

139. *Id.* at 856 *citing* Stuart M. Brooks et al., "The Spectrum of Irritant-Induced Asthma," *Chest* 113(1):42 (1998).

140. *Minner*, 791 A.2d at 857.

141. *Minner v. American Mortgage & Guaranty Co.*, 791 A.2d 826 (Del. Super. 2000).

142. *Id.* at 857.

143. *See, e.g., Schmaltz v. Norfolk & W. Ry.*, 878 F.Supp. 1119 (D.Ill. 1995) (The plaintiff's expert failed to come forward with sufficient empirical support for causation opinion that the plaintiff's condition, diagnosed as RADS, was caused by exposure to herbicides).

144. *Merrell Dow Pharms., Inc. v. Havner*, 953 S.W.2d 706, 714 (Tex. 1997).

145. Hutchinson, *supra* note 77, at 24 *citing Cavallo v. Star Enterprise*, 892 F.Supp. 756, 771 n. 34 (D.Va. 1995).

146. *Allison v. Fire Insurance Exchange*, 98 S.W.3d 227, 239 (Tex. App. 2002) citing Havner, 953 S.W.2d at 714-15.

147. *Allison v. Fire Insurance Exchange*, 98 S.W.3d 227 (Tex. App. 2002).

148. *Id.* at 239.

149. *Havner*, 953 S.W.2d at 717-23.

150. *Allison*, 98 S.W.3d at 239-40.

151. *Id.* at 240.

152. *Havner*, 953 S.W.2d at 714.

153. Hutchinson, *supra* note 77, at 25.

154. *Id.* at 26.

155. 278 F.Supp.2d 744 (D.Va. 2003).

156. *Id.* at 745–46.

157. *Id.* at 764.

158. *Id.*

159. *Id.*

160. *Dick v. Pacific Heights Townhouses*, 2002 WL 31117253 (Cal. App. 2002).

161. *Id.* at 2.

162. *Id.* at 3.

163. *Id.* at 9.

164. *Alberty v. J.C. Partners Limited*, 2000 WL 1807273 (Ohio App. 2000).

165. *Id.* at 1.

166. *Id.*

167. *Id.* at 3.

168. *Id.*

169. *Id.* at 4.

170. *Watters v. Department of Social Services*, 849 So.2d 724 (La. App. 2003).

171. *Id.* at 733.

172. *Id.* at 725.

173. *Id.*

174. *Id.* at 732–33.

175. *Id.* at 733.

176. *Id.*

177. *Id.* at 734.

178. *Graham v. Laurtec, Ltd.*, 2003 WL 23512133 at *2 (Mich. Cir. 2003)

179. *Id*. at *1.

180. *Id*. at *6.

181. *Id*.

182. 772 A.2d 792 (Del. 2001).

183. *Id.* at 800.

184. *Id.*

185. *Id.*

186. *Id.*

187. *Id*.

188. 2004 WL 3152375 (Mass. 2004)

189. *Id*. at *3-4.

190. *See* http://www.nyc.gov/html/doh/epi/moldrpt1.html.

191. *Id.*

192. *Id.*

193. *Id.*

194. *Id.*

195. *Id.* The American Industrial Hygiene Association has a listing of accredited laboratories available on their website at http://www.aiha.org.

196. *See* Frederick Fung, "What Is Mold AllAbout?" in: *Mold and Related Toxin Litigation and Claims Handling*, April, 2002 at 159.

197. Environmental Protection Agency, *Mold Remediation in Schools and Commercial Buildings, Appendix B: Introduction to Molds* (2001). *See also*, http://www.epa.gov/iaq/molds/appendb3.html.

198. ACOEM, *supra* note 49.

Part III

Chapter 3

Toxic Mold Law in Arizona

Dennis I. Wilenchik, Esq.

Synopsis

3.1 Introduction

While it may seem odd at first glance, mold has become a well-publicized phenomenon over the last few years in Arizona and other dry climate states, places where you might least expect it. A recent study conducted by University of Arizona researchers found mold in 64 percent of samples taken from homes in Tucson, Arizona.[1] Although only 0.2 percent of samples collected nationwide in the study contained the well-publicized toxic mold *Stachybotrys chartarum*, also known as "black mold," the study's findings indicate that a majority of homes have some form of mold within.[2] The study found that "[w]indowsills were the worst places for mold, testing positive 88 percent of the time. Refrigerator seals ranked second (84 percent), then areas below kitchen sinks (83 percent) and air vents (82 percent)."[3] Therefore, it would seem that mold is all around us, and that humans are continuously exposed to mold. Notwithstanding, exposure has been highly publicized almost to a frenzy.

As a result of a larger number of mold and water damage claims in Arizona, home insurance rates are increasing and insurers are beginning to exclude coverage. While Texas and California have had higher numbers of mold claims, insurers doing business in Arizona say that a high number of homeowners' claims related to mold and water damage have raised insurance rates.[4] "Of the 104 home-insurance companies currently writing new policies in Arizona, 92 posted rate increases [in 2001], and most of the increases exceeded 10 percent."[5] State Farm Insurance paid $11.6 million in Arizona in 2001 related to water damage alone.[6]

In addition, the State of Arizona has spent millions of dollars in an effort to clean up mold problems in government buildings.[7] The government spent roughly $5 million to assess and clean a Yuma, Arizona high school, built a new elementary school in Northeast Phoenix at a cost of $5.5 million because cleaning up the old building was cost-prohibitive, and replaced the U.S. Geographical Survey's field center in Flagstaff at a cost of $4.7 million.[8] The latter structure had roots to NASA's Apollo space program.[9] Other noteworthy Arizona buildings and structures, including theatres, airports, courthouses, and water treatment plants, will be cleaned or replaced because of severe mold problems.[10]

While mold has been around since the beginning of time, the notion of mold exposure litigation is a recent phenomenon. Large jury verdicts, including a $32 million jury verdict in Texas, have helped fuel the fire.[11] Television, print, and Internet coverage have also brought mold awareness into our consciousness. This chapter attempts to address the state of toxic mold law in Arizona.

3.2 Possible Explanations for Arizona's Mold Problems

Mold requires wet conditions to initiate and maintain growth. In determining the source of leaks and water intrusion, homeowners or inspectors should look at several places. Obviously, common suspects are inadequate or improper window or roof flashing and defective windows or surrounds. Blocked weep screed openings at the base of exterior stucco or around windows due to sunscreens are also typical places to find mold. Reducing moisture in a home may include fixing leaks and seepage, using exhaust fans in kitchens and bathrooms, limiting the use of humidifiers or swamp coolers, and using air conditioners.[12] While mold growth is impossible in dry conditions, many factors have increased the number of occurrences in Arizona homes, schools, and office buildings, including the pace of construction, the construction materials used, Arizona's extreme swing in temperatures, and homeowner maintenance or lack thereof.

A. The pace of construction in Arizona

Some people blame mold on Arizona's frantic pace of construction. Building permits in Phoenix in 2004 hit 60,873, up an astonishing 27.5% from 2003.[13] Factors contributing to the record number of building permits in the Phoenix area include low interest rates, people buying second homes or retirement homes, and the large number of people choosing to invest in real estate rather than the stock market. The large influx of residents into Arizona is another reason. In 1990, the Census Bureau estimated that there were 3,665,228 people living in Arizona.[14] Ten years later, in 2000, the Census Bureau estimated that there were 5,130,632 residents.[15] The difference represents an increase of nearly 40 percent and the Census Bureau expects continued population growth in Arizona at alarming rates.[16]

The frantic pace of construction has led some critics to claim that homes are poorly constructed. While the large majority of homes in Arizona are well built, given the government inspections required at critical stages, the fact is that homes remain one of the few manufactured items constructed outdoors and in uncontrolled conditions that vary widely. Construction quality does not vary greatly among homebuilders despite popular misconceptions, given the common subcontractor pool available. Constructing affordable houses fast is a key to success in an ultra-competitive market, but the demands of speed, affordability and quality aren't always compatible.[17]

Less-reputable builders will try to cut corners. In some instances, cutting corners could eventually lead to a mold problem. For example, in the past, some builders chose not to flash properly and wrap waterproofing paper properly around windows.[18] While the cost to install such waterproofing paper may have been only around $150 per house, builders faced an astronomical bill when problems arose after the fact.[19] Repair estimates can exceed over hundreds of thousands of dollars per home to repair defects including leaking windows and roofs, and various related stucco problems.[20]

Unskilled and untrained workers further compound construction problems. Overall, the construction industry in Arizona added an additional 8,100 jobs between November 2002 and November 2003.[21] Given the large number of new workers in the construction industry, it is hard to imagine that all are highly qualified. Even if a homebuilder uses waterproofing paper, if an untrained worker does not know how to properly install the paper, the house may later encounter problems.[22]

Roofing is another area where there is often shoddy or sloppy work, due to the fact that there are limited inspections, and the roof is not easily seen. In addition, lack of compaction of sewer and plumbing trenches and underground plumbing leaks may contribute to water intrusion.

B. Building materials used in Arizona

Any place where water intrudes is a possible source of mold growth. All it takes is twenty-four to forty-eight hours before mold begins to grow on building materials. Mold often enters a home through floodwaters, leaks, the outdoor environment (soil, dust, and dirt) through building materials at the time of construction, and through the misuse of toilets and showers. Therefore, homebuilders must build homes that manage moisture effectively.

The National Association of Homebuilders recommends that builders should (1) use quality building products, especially with respect to siding, shingles, windows, and pipes; (2) carefully install roofs and windows and make sure that roofs and windows are properly flashed; (3) reduce the possibility of condensation on cold surfaces by adding proper insulation; (4) make sure that rainwater drains away from the building through proper landscaping and adequate downspouts; (5) make sure that building materials like wallboard are dry at the time of installation; (6) make sure that HVAC systems are properly designed and working; and (7) make sure that moisture-generating activities like showers, dryers, and cooking appliances are vented to the outside.[23]

While the Western one-coat stucco system used through much of Arizona are popular with homebuilders, the system is not without its share of problems. Arizona's stucco system is typically designed to allow for moisture in hairline stucco cracks to drain to the outside environment over a membrane and down through to weep screeds that allow the water to evacuate. However, where moisture infiltrates the interior wall cavity through tears or voids in the underlying membrane and encounters wood, cellulose, or insulation fiber, various mold problems may arise in wall cavities. Insulation may reduce the ability of water to drain to the outside environment, reducing the rate at which the wall dries. Thin stucco systems can also crack excessively, allowing water to seep into the walls.[24]

Arizona homebuilders often use different materials than their counterparts in other portions of the country, making it difficult to readily compare various mold studies. In drier regions such as Arizona and Nevada, air conditioning is used more than evaporative coolers, walls are ventilated differently, and homebuilders use different insulation. Therefore, building interiors are generally cooler than the exterior for most of the year. In Phoenix and Tucson, vapor barriers in roofing are generally not necessary because stucco is a cladding system designed to absorb some moisture. Due to negative pressure, mold spores trapped in walls may be drawn throughout the home via the air conditioning.

Materials used may also contribute to the mold problem. For example, paper backed gypsum board, the use of insulation and wall-to-wall carpets may bring about mold growth.[25] Additionally, modern airtight construction may contribute to the problem. "Beginning in the 1970s, spurred by the energy crisis, architects started to design more airtight buildings that retain heat or air conditioning increasing energy efficiency. However, when airtight buildings with poor ventilation have water leaks, 'they trap high humidity and become hothouses for passing mold spores.'"[26] Homeowners renovating older homes can also create a more airtight environment by installing low-e windows, designed to keep out heat during the summer and in during the winter.

C. Arizona's warm temperatures

Arizona's warm temperatures, which incubate and help mold flourish, can also contribute to the growth of mold. Average monthly temperatures in Phoenix, Arizona range from 54 degrees Fahrenheit in January and December to 93 degrees Fahrenheit in July, and Phoenix's average high temperature reaches a scorching 104 degrees Fahrenheit in July.[27] Temperatures are also warm in Tucson, Arizona, where average monthly temperatures range from 52 degrees Fahrenheit in January and December to 87 degrees Fahrenheit in July.[28] In addition, Arizona's "dry heat" may contribute to the growth of mold. Phoenix averages just 8.29 inches of precipitation a year.[29] While mold is not limited to warm climates, these temperatures might add to the growth of mold in Arizona.

Because of Arizona's warm temperatures, governmental and regional standards regarding mold from other climate zones may not be relevant. For example, New York has a completely different climate than Arizona, so the well-publicized New York City "Guidelines on Assessment and Remediation of Fungi in Indoor Environments"[30] should be limited to New York City or at most, the Northeastern portion of the United States. Sanctioned guidelines specific for Arizona are not currently available, and many mold remediation specialists choose to create their own protocol.

D. Homeowners' contribution to the mold problem

In addition, homeowners can contribute to mold intrusion and growth. This is especially true if a homeowner fails to undertake maintenance such as caulking and filling cracks. In addition, alterations to grading and drainage plans and over-watering of lands may cause excess moisture to develop under slabs and migrate through slab cracks. Certain homebuilders take it upon themselves to educate homeowners about such proper maintenance. "By providing homeowners with information about the relationship between proper maintenance and mold growth, homeowners will better understand their environment and their own obligations. An information disclosure statement should be given to homeowners before the sale and should be incorporated by reference into each sales agreement."[31] The information disclosure statement should include the following information: (1) mold growth is possible even where there is no moisture, (2) potential moisture sources include water in the air-conditioning unit and refrigerator drip pans, showers, bathtubs, sinks, and pipe leaks, (3) controlling and preventing moisture is possible if leaking pipes are promptly fixed and sinks and bathtubs are checked for standing water or signs of mold, and (4) a waiver of liability for mold growth caused by the homeowner's failure to follow the builder's recommendations.[32]

Homeowners often create mold growth situations by introducing excessive water to the stem walls. In Arizona, this typically plays out in several ways. First, homeowners may introduce excessive water to the stem walls by disrupting or disturbing drainage. In designing homes, homebuilders routinely engineer drainage troughs designed to carry water, especially during a rainstorm. Arizona homebuilders will provide the original homeowners with warnings not to disturb those features. Nevertheless, homeowners will often disturb these troughs by installing swimming pools and landscape features. Large lot spaces and the popularity of swimming pools contribute to a situation where engineered drainage and grading plans are sometimes disturbed, which causes water to run towards or pool up around a home. Second, homeowners may plant vegetation too close to a home, which might introduce excessive water to a stem wall. Placing plants and trees next to a home may cause the stem walls and foundation to be subject to an excessive amount of water from sprinkler and irrigation systems. If a home is not designed to handle this excessive amount of water, the homeowner may face mold growth.

Under Arizona law, a plaintiff generally is required to mitigate damages, and to not voluntarily incur them. Specifically, a plaintiff cannot recover damages that a plaintiff should have foreseen or that he could have avoided by reasonable effort without undue risk, expense, or humiliation.[33] For example, a homeowner should have to accept an offer by the homebuilder to address water intrusion. Homeowners should also be required to cease activities noted in the previous paragraph, as well as fix known leaks.

3.3 Typical Claims Asserted by Plaintiffs in Arizona

Mold litigation is a rather new phenomenon and there is not much reported case law in Arizona or other states on it beyond insurance coverage issues. Plaintiffs will typically assert claims based on both tort law (negligence) and contract law (warranty) when filing an action. Often, plaintiffs will attempt to assert other claims, such as strict liability and fraud, but these claims have had little success in Arizona.

A. Introduction

Before discussing possible causes of action, it is imperative to review the Statute of Repose, A.R.S. § 12-552, which sets a limit on the time a plaintiff may bring an action against "a person who develops or develops and sells real property, or performs or furnishes the design, specifications, surveying, planning, supervision, testing, construction or observation of construction of an improvement to real property" to eight years from the date of substantial completion.[34] If, however, the injury occurred during the eighth year, or was not discovered until the eighth year, a plaintiff has an additional year in which to bring the action.[35] Under no circumstances may an action commence when more than nine years has passed since substantial completion of the structure. The statute applies to claims that are contract-based, such as those arising under an express warranty or of the workmanship and habitability, but not to personal injury or death actions.[36] In addition, the Statute of Repose does not shorten a separate warranty provision in an express warranty.[37]

The eight-year period commences at the substantial completion of the building. The substantial completion date is the first of either (1) the time when the property is first used by the owner or occupant; (2) the date the property is first available for use according to the contract or agreement; or (3) the time of the final inspection by a governmental body.[38]

There is no debate whether the separate breach of contract limitation period set forth in A.R.S. § 12-548 applies to implied warranty claims so as to trump the Statute of Repose. The Statute of Repose makes it clear that it applies regardless of any other statute of limitations. An area of unresolved concern is whether a repair made under the express or implied warranty establishes a new base for the running of a breach of contract or negligence statute of limitation on the repair that may exceed or extend the Statute of Repose. This problem frequently arises in settlement negotiations in older homes approaching the period of repose. As to the limitation period for implied warranty claims, it would seem that an implied warranty is a claim that is based in contract, although a creation of law, and therefore may be governed by the six-year limitation.[39] There is no known authority other than Registrar of Contractor Rules that might apply that extends the Statute of Repose by a repair.

Regarding existing structures, buyers and sellers of homes usually have no special duties to one another, and their interests are by nature conflicting and adverse.[40] Sellers must disclose latent defects that they are aware of, but it is often difficult to prove knowledge.[41] The seller's common law duty to disclose defects generally applies only to matters uniquely within the seller's knowledge that could not be known by the buyer's exercise of reasonable due diligence.[42] The interests of a seller's broker are also in conflict with the buyer's interests, as they must exercise reasonable due care and diligence to effect a sale to the principal's best advantage.[43] Brokers must also disclose only to their clients information they possess pertaining to the transaction involved.[44]

B. Negligence

Claims made in mold cases include negligence. These relate to claims of personal injury dealing with anything from rhinisitus to aggravation of asthma to lung disorders. Claims that furniture and belongings must be destroyed due to contact with mold are also prevalent. For a plaintiff to succeed on a negligence claim, he must prove four elements: duty, breach of duty, causation, and damages.

In proving breach of duty, a plaintiff is required to show a standard of care violated by the builder through the testimony of another builder. Testimony from an engineer or architect will not suffice.[45] It would be prudent to look into applicable building codes, plans, and specifications to find deviations. In Arizona, homebuilders typically hire subcontractors to construct homes. Thus, mold injury claims generally involve active negligence by a subcontractor and not the homebuilder. Plaintiffs must allege that builders are liable for negligent independent selection, hiring, and supervision of subcontractors, who are alleged to have actually caused the mold problem by negligent construction. Plaintiffs who claim that the homebuilder or general contractor negligently supervised its subcontractors must introduce testimony from an expert homebuilder to establish the requisite standard of care.[46] However, as noted below, general contractors are commonly not liable for negligent acts of their independent subcontractors.

Causation is a problem for plaintiffs to prove. Plaintiffs must show that the defendant's failure to follow the applicable standard of care proximately resulted in injury. Because there is little established support showing a link between exposure to even allegedly toxic molds like *Stachybotrys chartarum* and permanent injury, plaintiffs will have a difficult time proving causation in toxic mold cases. Even assuming that a plaintiff has damaged personal property, the burden still remains on the plaintiff to prove the defendant's negligence.[47]

The Arizona Supreme Court held in *Woodward v. Chirco Construction Co.*[48] that plaintiffs can bring an action in both contract and tort law for damages related to "defects" in construction.[49] The Supreme Court explained the difference between contract and tort claims by announcing that "if a fireplace collapses, the purchaser can sue in contract for the cost of remedying the structural defects and sue in tort for damage to personal property or personal injury caused by the collapse. Each claim will stand or fail on its own; a distinct statute of limitation applies to each."[50]

Plaintiffs bringing suit against homebuilders in negligence must therefore claim either personal injury or personal property damage. In *Nastri v. Wood Brothers Homes, Inc.*[51], the Arizona Court of Appeals decided that a plaintiff could not maintain an action under a negligence theory against a homebuilder without property damage or personal injury.[52] In that case, the plaintiff's tort claim failed because the damage in the case involved structural defects to the property; plaintiff did not have any personal injury or damage to personal property.[53]

In *Hayden Business Condominium Association v. Pegasus Development Corp.*, 105 P.3d 157 (Ariz. App. Div. 1 2005), the court clearly defined damages to the structure by stating: "carpeting and landscaping are not considered personal property separate from the property. Carpeting is a fixture and part of the building's structure. *See In re Bernhardt*, 186 B.R. 889, 891 (Bankr.E.D.Pa. 1995) (wall-to-wall carpeting is generally considered a fixture). Re-carpeting and re-landscaping are all part of remedying defective work." Thus, Arizona law is clear that, in order to bring suit under an implied warranty theory, discussed *infra*, there must be damage to the structure of the house. If a homeowner intends to bring suit in negligence, he or she must show personal injury or property damage separate from the structure of the house.

Arizona homebuilders are ordinarily general contractors, and construct homes through the use of various specialist subcontractors. Under the Restatement of Agency, subcontractors retained by a general contractor are "independent contractors."[54] In *Wiggs v. City of Phoenix*[55] the Arizona Supreme Court held that while an employer is liable for the negligence of its employees under the doctrine of *respondeat superior*, an employer is usually not liable for the negligence of an independent contractor.[56] In the *Wiggs* ruling, the Arizona's Supreme Court expressly noted that Arizona follows the general rule that a general contractor is not liable for the acts of an independent contractor, in the absence of a recognized exception to that general rule.[57] The Supreme Court found that the City of Phoenix *could* be liable for the torts of an independent contractor related to work on a public roadway, because the city had a "non-delegable duty" to maintain the roadway.[58] The "non-delegable duty" exception is recognized in the Restatement of Torts, but neither the Restatement nor Arizona statutes nor case law have found that the duties of a general contractor constructing a home cannot properly be delegated to subcontractors.

Courts in Arizona and in other jurisdictions have found non-delegable duties in "inherently dangerous" activities. For example, in *Ft. Lowell-NSS Ltd. Partnership v. Kelly*, 166 Ariz. 96, 800 P.2d 962 (1990), the Arizona Supreme Court found a landowner liable for injuries to the employee of an independent contractor who received severe electrical shock while demonstrating an electronic security system. Additionally, the mere existence of a contract between a general contractor and a subcontractor does not create a non-delegable duty.[59] Finally, in explaining the non-delegable exception, courts in other jurisdictions have held that even where a house fire resulted in death, a general contractor or employer is still not liable for the negligent acts of its subcontractors.[60]

C. Implied warranty of habitability

Under the implied warranty of habitability, a home must be reasonably suited for its intended use.[61] The implied warranty of habitability and fitness of a house arises from the contractual relationship between the homebuilder and purchaser, although it is a creation of law based on public policy.[62] The purpose of the implied warranty of habitability and workmanlike performance is to rectify latent and defective design, workmanship or construction within the Statute of Repose, A.R.S. § 12-522.[63] The implied warranty of habitability and workmanlike performance was judicially created to ensure that construction is done in a workmanlike manner, and the implied warranty provides homeowners, including those who do not have privity of contract with the homebuilder, with an important guarantee that their homes are habitable.[64]

In *Nastri*, the Arizona Court of Appeals found that the attempted disclaimer of the implied warranty of habitability was void against public policy as to innocent subsequent purchasers.[65] The court did not comment on a knowing and fully agreed-upon waiver. When the plaintiff is a subsequent purchaser rather than the original homeowner, the burden is placed on the subsequent homeowner to demonstrate that the builder-vendor caused the defect and that the action was brought within the appropriate timeframe prescribed by A.R.S. § 12-552.[66] In addition, the plaintiff must show not only that the defect originated with the builder-vendor, but also that the defect was latent, as discussed below. The builder-vendor can claim that the defects are not attributable to him, that the defects are the result of ordinary wear and tear or age, that the plaintiff discovered the defects within the applicable statute of limitations, and that the plaintiff made substantial changes to the building that resulted in the problem.[67]

The implied warranty of habitability only applies to "latent defects which become manifest after the subsequent owner's purchase and which were not discoverable had a reasonable inspection of the structure been made prior to purchase."[68] In other words, the implied warranty of habitability does not apply where the purchaser could have discovered the defect through reasonable inspection.[69] Regarding latent defects, if the seller of real property fails to disclose a defect known to the seller or fails to give an appropriate opportunity to discover the facts, the doctrine of *caveat emptor* no longer applies and the vendor is subject to tort liability for nondisclosure.[70] Even though a potential home purchaser is afforded a certain number of days to conduct an inspection, the seller cannot withhold information about the presence of mold if the mold is a latent defect. The standard form contract used by realtors affords the buyer a certain number of days to conduct any type of inspection he desires. If a home inspector fails to find mold during the inspection and mold is present, the home purchaser may have an action against the home inspector.

The implied warranty of habitability is limited by the reasonable lifespan of the item in question. In the past, Arizona courts have found that twelve years was not an unreasonable period of time for an implied warranty of habitability and workmanship to exist for stucco.[71] The Arizona Legislature responded in 1989 by enacting the Statute of Repose (A.R.S. § 12-552), which prevents plaintiffs from bringing any contract-based claims arising from the construction of a home more than eight years (or nine years) after "substantial completion." As noted above, A.R.S. § 12-552 sets a bright line rule that all construction defect claims must be brought either eight or nine years after the substantial completion of the home, no matter the "reasonable lifespan" of an item. For example, even if stucco is designed to last between thirty and fifty years, A.R.S. § 12-552 limits a construction defect claim to either eight or nine years, depending on when the homeowner discovered the defect.

In discussing mold claims in particular, some critics point out that "water intrusion is likely to occur because of defects in housing components that should last longer than eight years—components such as the roof, plumbing and stucco. The damage from defects in these components could occur well after the Statute of Limitations expires, but years before the end of the expected lifespan of the components. Nevertheless, homeowners cannot recover under the implied warranty in such situations."[72] A.R.S. § 12-552 properly acts to prevent homebuilders from being indefinitely liable for the failure of components in a home. Homebuilders are no longer eternal insurers of a home.

Note that actions brought under the implied warranty of habitability arise out of a contract and therefore fall within A.R.S. § 12-341.01, providing for attorneys' fees and costs.[73]

D. Other potential causes of action and homeowner complaints

Arizona does not recognize strict liability or product liability in the context of homebuilding.[74] Strict liability and product liability are claims that do not require fault on the part of the defendant, and are typically reserved for hazardous and dangerous activities, manufacturing product defects, and product design defects. While the seller of a "product" in a defective condition is usually liable in strict liability, policy reasons do not justify characterizing homes or homebuilding materials as "products."[75]

In addition, no Arizona court has addressed damages for emotional distress or lost wages in a construction defect or mold case, although if physical injury could be established in a negligence case, it is theoretically possible. Other jurisdictions considering such claims have denied recovery.[76] In Arizona, the measure of damages for a construction defect claim is either the cost of repair or, in the situation where repairs are economically wasteful, the diminution in value. The measure of damages does not contemplate damages for emotional distress.[77] In any event, allegations of emotional and mental distress are not compensable unless based on some continuing, non-temporal physical injury, which might be difficult for a plaintiff to prove.[78] If a homebuilder is faced with an emotional distress claim, they should argue that contracts for the purchase of homes or real property are not considered the type of contract that would cause emotional distress as a foreseeable damage from its breach.[79]

Homeowners are also unable to recover "stigma" damages related to their home. Homeowners often complain that once they have encountered a mold problem, they are unable to sell their home for full value, or even at all. A plaintiff should not be entitled to "stigma" damages if a home is truly remediated and certified to be free of mold. Arizona case law has long established that the cost to repair is the proper measure of damage, unless the repair is economically wasteful or involves complete destruction.[80] Diminution in value has only been recognized in Arizona as an issue when permanent and calculable physical damage to chattels has occurred.[81] Furthermore, diminution in value is only applicable when physical damage has occurred, and the relief to be awarded is the cost of repair or replacement not to exceed the amount of diminution in value.[82] Stigma damages are not compensable because "[a]n award of damages cannot be based on conjecture or speculation."[83] In any event, most new home construction contracts in Arizona have express limitations on the measure of damages contemplated, and include repair or replacement as the sole agreed-upon measure of damages.

In other jurisdictions, plaintiffs have brought actions based on improper appraisal,[84] breach of contract,[85] breach of duty of good faith and fair dealing,[86] civil rights violations,[87] deceptive trade practices,[88] fraud,[89] consumer fraud,[90] infliction of emotional distress,[91] and libel and slander.[92] Whether some or all of these potential causes of action would stand in Arizona is certainly debatable.

3.4 Recent Mold-Related Legislation in Arizona

The recent attention given mold by the media has prompted the Arizona Legislature to introduce bills regarding mold and other construction lawsuit issues. Although not all of the bills have been signed into law, as long as mold claims are popular, more bills will likely be introduced and signed into law.

A. SB 1432: Residential and Commercial Mold; Study Committee (2002)

While states like California have recently passed bills regarding mold or other related pollutants,[93] Arizona has not yet adopted such a law. The Arizona State Legislature, however, did introduce a bill during the 45th Legislature in 2002, which ultimately was held up in the House. Senate Bill 1432 established a Joint Legislative Study Committee on Residential and Commercial Mold Contamination.[94] The committee was to be comprised of numerous home building professionals, engineers, legislators, and homeowners.[95] This committee was to study the financial, environmental, and health effects associated with the presence of indoor mold in residential and commercial buildings, and was to submit a report, as well as policy changes or recommendations, by December 1, 2002.[96] Although the end result was that the bill was held up, future representatives will undoubtedly introduce bills similar in nature in upcoming sessions.

B. HB 2620 (2002) and SB 1311 (2004): Purchaser Dwelling Action; Opportunity to Repair

Even though Senate Bill 1432 noted *supra* was not signed into law, the Arizona Legislature has passed other significant laws in recent years. In particular, Arizona House Bill 2620[97] amended A.R.S. § 33-2003 and introduced A.R.S. § 12-1361 *et seq.* These statutes are important because the homebuilder or seller now has the statutory right to make an offer to repair a property before the purchaser can file a "dwelling action" or else there are severe consequences.

"Dwelling actions" are defined as "any action brought by a purchaser against the seller of a dwelling arising out of or related to the design, construction, condition or sale of the dwelling" and would encompass mold actions.[98] The heart of House Bill 2620's drastic changes can be found in A.R.S. 12-1363, entitled "notice and opportunity to repair." Purchasers must provide the seller the basis of the action by certified mail ninety days before filing the action.[99] However, A.R.S. § 12-1361 *et seq.* do not apply where the alleged defects involve "an immediate threat to the life or safety of persons occupying or visiting the dwelling."[100]

After the seller receives notice, the seller is able to inspect the dwelling unit or units to determine the nature and cause of the alleged defects and to assess possible repairs and replacements.[101] This inspection period provides the seller an opportunity to inspect the property for mold. Naturally, the purchaser must make his dwelling available to the seller for inspection, and the statute provides that the purchaser must comply within ten days after the seller makes a request for inspection.[102] In inspecting the unit, the seller is allowed to use "reasonable measures, including testing, to determine the nature and cause of the alleged defects and the nature and extent of any repairs or replacements necessary to remedy the alleged defects."[103] The seller is also required to return the premises to their pre-testing condition.[104] In this situation, a seller may want to bring a mold specialist or remediator to the premises in order to thoroughly investigate the claim.

Sixty days after the purchaser provides notice, the seller must respond to the purchaser's notice by certified mail.[105] In his response, the seller may offer to repair or replace the defects, to have the alleged defects repaired or replaced at the seller's expense, or provide the purchaser monetary compensation.[106] If the seller offers a solution to the problem, the seller must describe in detail what he plans to do and provide a reasonable completion date.[107] A mold remediator may wish to describe his services at this time. In the case of a single-unit dwelling action, where the complaint has not yet been filed, if the seller does not respond to the purchaser's notice within sixty days, the purchaser can file his action without waiting for the expiration of the ninety-day period.[108]

Within twenty days after receipt of the seller's response, the purchaser must respond by either accepting or rejecting the seller's offer by certified mail.[109] The purchaser may also include a counteroffer.[110] If the purchaser rejects the seller's offer, he must state the basis for his rejection.[111] The seller then has ten days within which to make his best and final offer by certified mail.[112]

In the event that the purchaser files an action, the statute provides that certain facts are inadmissible in court, including the purchaser's notice and all good-faith responses, counteroffers and final offers made by the purchaser and seller.[113] Other relevant provisions of A.R.S. § 12-1363 include allowing the purchaser to amend the original notice[114] and the tolling of the Statute of Limitations.[115]

A.R.S. § 12-1364 provides that "[i]n any contested dwelling action, the court shall award the successful party reasonable attorney fees, reasonable expert witness fees and taxable costs." A.R.S. § 12-1364 also incorporates a version of Rule 68 of the Arizona Rules of Civil Procedure regarding offers of judgment. If the purchaser rejects the seller's best offer and the final judgment is even less favorable to the purchaser than the seller's offer, the court deems that the seller is the successful party and is entitled to attorney fees, expert fees, and taxable costs.[116] Under A.R.S. § 12-1365, written contracts for the sale of newly constructed homes must provide specific language stating that the purchaser has the right to file an action with the Arizona Registrar of Contractors. Finally, A.R.S. § 12-1361 *et seq.* do not apply to personal injury claims, claims for damage to property other than a dwelling, common law fraud claims, to claims where the sale contract contains commercially reasonable alternative dispute resolution procedures, and to claims solely seeking recovery of monies expended for repairs to alleged defects already repaired by the purchaser.[117]

3.5 Reported Mold Case Law in Arizona

The only Arizona cases discussing mold damage revolve around insurance issues. In *Cooper v. American Family Mutual Insurance Co.*,[118] decided by the Federal District Court in Arizona, the insured brought an action against the insurance carrier to recover for mold damage caused by a plumbing leak, which also damaged drywall and flooring.[119] The court, on cross-motions for summary judgment, held that the homeowner's insurance policy did not cover mold damage.[120] The relevant provision in the insurance policy stated that mold loss was not covered "regardless of any other cause or event contributing concurrently or in any sequence to the loss."[121] Another provision of the policy denied coverage for "smog, rust, corrosion, frost, condensation, *mold*, wet or dry rot."[122]

In *Liristis v. American Family Mutual Insurance Co.*[123] the Arizona Court of Appeals held that the insurance policy *covered* mold damage, including the cost of removal or repair of the damage. Interestingly enough, the defendant was the same as in the *Cooper* case noted above, and the insured had the same insurance exclusions.[124] In *Liristis*, the plaintiffs claimed that a fire (and the water used to extinguish the fire) caused their mold problems.[125] The court found that the insurance policy distinguished between (1) mold damage and (2) loss caused by mold, and did not exclude mold damage caused by a covered event like fire.[126] If the mold resulted from the fire, then the cost of removing the mold was not a loss separate from or caused by the mold itself.[127] In reversing the granting of plaintiff's motion for summary judgment, the court remanded back to the trial court to determine whether some or all of the mold damage was caused by the fire and the water used to extinguish it.[128] It is difficult to distinguish between *Cooper* and *Liristis* other than to say that they were decided by different courts.

3.6 Admissibility of Expert Witness Testimony in Arizona

State courts in Arizona follow the *Frye* standard regarding the admissibility of scientific evidence. Under the *Frye* standard, the court considers whether the methodology employed by an expert is "generally accepted" within the scientific community.[129] Arizona's courts have made certain limited exceptions to the *Frye* standard "when a qualified witness offers relevant testimony or conclusions based on experience and observation about human behavior for the purpose of explaining that behavior."[130] This exception is further limited to opinions which are based on a qualified expert's own research, observations, and experience, as well as the research of others with which he is familiar.[131] Therefore, pursuant to *Frye*, an expert relying upon others' theories and methods will need to show that those theories have been generally accepted in the scientific community.

In Arizona, *Frye* hearings are required before the admission of expert testimony that relies on new or novel scientific tests or techniques.[132] *Frye* hearings are only necessary when an opposing party makes a timely request supported by evidence that the community does not generally accept an expert's technique or method. Thereafter, expert testimony is admissible only if the proponent of the expert testimony can demonstrate that the expert's methodology is generally accepted in his particular field.[133] *Frye* hearings are often conducted before trial in order to timely resolve the general acceptance issue.[134] Although there are no Arizona *Frye* cases dealing with the admissibility of expert evidence regarding mold, Arizona courts have applied *Frye* to a number of other areas.[135]

It should be successfully argued that expert testimony regarding mold must pass the *Frye* standard in order to be admissible, particularly as to its effect on the health of individuals. The Arizona Supreme Court has stated *in dicta* that an expert's opinion, which is nothing more than *ipse dixit* or an unsupported assertion, should not be allowed.[136]

> The court's obligation to exclude unproven and potentially misleading scientific evidence is 'especially sensitive in cases where the plaintiff claims that exposure to a toxic substance caused his injury, [because a] jury may blindly accept an expert's opinion that conforms with their underlying fears of toxic substances without carefully understanding or examining the basis for that opinion.'[137]

Rule 702 of the Arizona Rules of Evidence allows an expert to testify regarding scientific, technical, or other specialized knowledge if it "will assist the trier of fact to understand the evidence or to determine a fact in issue." In toxic mold cases, the trial court should properly exclude expert testimony that is unproven or misleading because it will not assist the trier of fact but rather will cloud the trier of fact's judgment. Specifically, the court should exclude expert testimony regarding a causal link between exposure to toxic mold and human health consequences because recent scientific publications have "concluded that indoor airborne levels of microorganisms are only weakly correlated with human disease or building-related symptoms and that a causal relationship has not been established between these complaints and indoor exposures to [*Stachybotrys*] *chartarum*."[138]

3.7 Conclusion

Few would dispute that a mold problem exists today, particularly in states with warmer climates like Arizona. Even though mold has existed since the beginning of time, only recently has it been brought to our attention through the mainstream media. While we may not know the full consequences of exposure to mold, particularly in the long run, one thing is for certain: mold claims are becoming ever more prevalent. For many people, their homes represent their largest investment. Unfortunately, many homeowners are disappointed once they discover that their dream home has a mold problem, and they quickly look to the homebuilder for relief. However, homeowners are also responsible for their mold problems by not properly maintaining their homes, by constructing swimming pools and tracking water into the home, or by adding vegetation or water features. In any event, the state of the science regarding mold is evolving quickly. No doubt, a response by the Arizona Legislature and the Arizona courts will follow and shape the future of mold law in Arizona over the next several years. The Arizona Legislature has already acted to pass laws related to construction defects, and the Arizona courts have also acted recently by publishing two decisions regarding insurance issues related to mold coverage. While we do not know exactly what the future holds for toxic mold law in Arizona, we do know that this problem will not quietly disappear. As long as people discover mold in their homes, they will file lawsuits in Arizona courts.

Endnotes

1. Kerry Fehr-Snyder, "Sneeze and Wheeze: Molds Make Themselves Right at Home," *Arizona Republic*, March 10, 2003, at A1.

2. Kelly A. Reynolds, Ph.D., research microbiologist, University of Arizona, Tucson; Jordan S. Josephson, M.D., director, New York Nasal and Sinus Center, and attending physician, Lenox Hill Hospital, New York City; March 10, 2003, presentation, American Academy of Allergy, Asthma and Immunology annual meeting, Denver.

3. Fehr-Snyder, *supra* note 1.

4. Catherine Burrough, "Insurance Premiums for Homeowners Rising," *Arizona Republic*, October 26, 2003, at H10; Christine L. Romero, "Arizona Insurance Rates for Autos, Homes Soar," *Arizona Republic*, March 27, 2002 at A1.

5. Burrough, *supra* note 4.

6. Romero, *supra* note 4.

7. Kerry Fehr-Snyder, "Mold Removal Costs Soar," *Arizona Republic*, July 27 2001, at A1.

8. *Id.*

9. *Id.*

10. *Id.*

11. Ballard v. Fire Ins. Exchange, 2001 WL 883550 (Tex. Dist. Ct. 2001) *aff'd in part, rev'd in part sub nom.* Allison v. Fire Ins. Exchange, 98 S.W.3d 227 (Tex. App. 2002) (affirmed $4,006,320.72 of the $32 million award and remanded the case for further calculations).

12. Star Lawrence, "Researchers Divided on Toxic Mold," available at http://my.webmd.com/content/Article/73/89002,htm?printing=true.

13. Glen Creno *et al.*, "Home building, prices soarded in 2004," Arizona Republic, January 27, 2005.

14. U.S. Census Bureau data for 1990, available at http://factfinder.census.gov/servlet/BasicFactsTable?_lang=en&_vt_name=DEC_1990_STF1_DP1&_geo_ id=04000US04.

15. U.S. Census Bureau data for 2000, available at http://factfinder.census.gov/servlet/BasicFactsTable?_lang=en&_vt_name=DEC_2000_SF1_U_DP1&_geo_id=04000US04.

16. By 2025, the Census expects Arizona to be the seveteenth most populous state with a population of 6.4 million people. U.S. Census Bureau data available at http://www.census.gov/population/projections/state/9525rank/azprsrel.txt.

17. Craig Harris et al., "Construction Quality Varies Widely," *Arizona Republic*, November 20, 2001, at A1.

18. *Id.*

19. *Id.*

20. *Id.*

21. Jonathan J. Higuera, "Jobless Rate in State Falls as Economy Makes Gains," *Arizona Republic*, December 19, 2003, at D1.

22. Harris et al., *supra* note 18.

23. National Association of Homebuilders, *A Builder's Guide to Handling Mold Claims and Litigation* (2001), page 3.

24. Harris et al., *supra* note 18.

25. Joseph Lstiburek et al., *Mold: Causes, Health Effects and Clean-Up*, (2002) page 2, available at http://www.eere.energy.gov/buildings/building_ america/pdfs/db/35012.pdf.

26. Thelma Jarman-Felstiner, "Mold is Gold: But, Will It Be the Next Asbestos?" *Pepp. L. Rev.* 30:529, 535 (2003) citing Andy Dworkin, "Mold Leaves a Trail of Victims, Lawsuits," *Oregonian*, August 19, 2001, at A1, available at 2001 WL3611638. *But see*, "Blame Moisture, Not Airtight Construction," *Sacramento Business Journa*l, July 15, 2002, available at http://www.bizjournals.com/sacramento/stories/2002/07/15/editorial3.html?t=printable (Moisture problems are related to poor construction of building components and the failure of a building system. Airtight buildings will prevent the infiltration of outdoor mold particles, but will trap mold contaminants if an indoor mold source exists).

27. http://www.weather.com/outlook/travel/climatology/monthly/USAZ0166?from=search.

28. http://www.weather.com/outlook/travel/climatology/monthly/USAZ0247?from=search.

29. http://www.weather.com/outlook/travel/climatology/monthly/USAZ0166?from=search.

30. These guidelines are available at http://www.ci.nyc.ny.us/html/doh/html/epi/moldrpt1.html.

31. National Association of Homebuilders, *supra* note 24, at 4.

32. *Id.*

33. *See, Coury Bros. Ranches, Inc. v. Ellsworth*, 103 Ariz. 515, 518–520, 446 P.2d 458, 461-463 (1968) (When pasture renter became aware of the fact that certain acreage contracted for would not be available for pasturing, it became the renter's duty to make a reasonable effort to prevent losses).

34. A.R.S. § 12-552(A).

35. A.R.S. § 12-552(B).

36. A.R.S. § 12-552(C), (D). Note that there is an open issue as to the applicability of A.R.S. § 12-552 to indemnity claims by a general contractor against its subcontractors.

37. A.R.S. § 12-552(D).

38. A.R.S. § 12-552(E).

39. *Woodward v. Chirco Constr. Co., Inc.*, 141 Ariz. 514, 516, 687 P.2d 1269, 1271 (1984).

40. *Oldenburger v. Del Webb Devel. Co.*, 159 Ariz. 129, 133, 765 P.2d 531, 535 (App. 1988).

41. *S. Develop. Corp. v. Pima Capital Mgnt. Co.*, 201 Ariz. 10, 17, 31 P.3d 123, 130 (App. 2001).

42. *Hill v. Jones*, 151 Ariz. 81, 85, 725 P.2d 1115, 1119 (App. 1986).

43. *Haldiman v. Gosnell Devel. Corp.*, 155 Ariz. 585, 589, 748 P.2d 1209, 1213 (App. 1987); *Vivian Arnold Realty Co. v. McCormick*, 19 Ariz. App. 289, 293, 506 P.2d 1074, 1078 (1973).

44. *Jennings v. Lee*, 105 Ariz. 167, 173, 461 P.2d 161, 167 (1969).

45. *See, e.g., Woodward v. Chirco Constr. Co., Inc.*, 141 Ariz. 520, 521–23, 687 P.2d 1276-78 (App. 1984) (trial court properly dismissed homeowners' negligence claim since no construction expert testified it was standard practice in the community to obtain a soil test prior to new construction).

46. *See id.*

47. *Payne v. M. Greenberg Constr.*, 130 Ariz. 338, 347, 636 P.2d 116, 125 (App. 1981).

48. *Woodward v. Chirco Constr. Co., Inc.*, 141 Ariz. 514, 687 P.2d 1269 (1984).

49. *Woodward*, 141 Ariz. at 516, 687 P.2d at 1271.

50. *Id.*

51. *Nastri v. Wood Bros. Homes, Inc.*, 142 Ariz. 439, 690 P.2d 158 (App. 1984).

52. *Nastri*, 142 Ariz. at 444-45, 690 P.2d at 163-64.

53. *Id.*

54. The Restatement (Second) of Agency § 220 delineates when workers are independent contractors. In pertinent part, § 220 states: (2) In determining whether one acting for another is a servant or an independent contractor, the following matters of fact, among others are considered:

(a) the extent of control, which, by the agreement, the master may exercise over the details of the work;
(b) whether or not the one employed is engaged in a distinct occupation or business;
(c) the kind of occupation, with reference to whether, in the locality, the work is usually done under the direction of the employer or by a specialist without supervision;
(d) the skill required in the particular occupation;
(e) whether the employer or the workman supplies the instrumentalities, tools, and the place of work for the person doing the work;
(f) the length of time for which the person is employed;
(g) the method of payment, whether by the time or by the job;
(h) whether or not the work is a part of the regular business of the employer;
(i) whether or not the parties believe they are creating the relation of master and servant; and
(j) whether the principle is or is not in business.

55. *Wiggs v. City of Phoenix*, 198 Ariz. 367, 10 P.3d 625 (2000).

56. *Wiggs*, 198 Ariz. at 369, 10 P.3d at 627.

57. *Id.*

58. *Wiggs*, 198 Ariz. at 369-70, 10 P.3d at 627–28.

59. *See e.g., Carrasquillo v. Holiday Carpet Service, Inc.*, 615 So.2d 862, 863 (Fla. App. 1993) ("the mere existence of a contract does not create vicarious liability as to third persons for the negligent acts of the independent contractor. There *must* be a nondelegable duty").

60. In *McGinnis v. Jim Walter Homes, Inc.*, 800 So.2d 140 (Ala. 2001), homeowners brought a negligence action against the contractor, electrical subcontractor, and electrical supplier seeking damages resulting from a fire in their home that killed their daughter. Citing to *Knight v. Burns, Kirkley & Williams Constr. Co.*, 331 So. 2d 651, 655 (Ala. 1976), the Alabama Supreme Court held that a general contractor is not liable for the alleged negligence of an independent contractor. *Id.* at 146. *See also, Rivera v. Goodenough*, 162 P.2d 498 (Cal. App. 1945) (a general contractor building a warehouse was not liable for the negligent acts of subcontractor installing trusses to support roof of the warehouse); *Hall v. Richardson Homes, Inc.*, 309 S.E.2d 825, 829 (Ga. App. 1983) (any negligence on the part of the subcontractor was not imputable to the homebuilder because the unrebutted evidence showed that septic tank system was installed by an independent subcontractor); *Miltz v. Borroughs-Shelving*, 497 A.2d 516 (N.J. App. 1985) (the general contractor was not liable for subcontractor's negligence in the installation of stairs, as there was no proof that the general contractor engaged an incompetent subcontractor or that installing the stairs was an inherently dangerous activity).

61. *Nastri v. Wood Bros. Homes, Inc.*, 142 Ariz. 439, 444, 690 P.2d 158, 163 (App. 1984) ("Rather, we find the test … that the home be reasonably suited for its intended use consistent with the Arizona decisions").

62. *Woodward v. Chirco Constr. Co., Inc.*, 141 Ariz. 514, 516, 687 P.2d 1269, 1271 (1984). *See also, Dillig v. Fisher*, 142 Ariz. 47, 50, 688 P.2d 693, 696 (App. 1984) (the rationale underlying the implied warranty of habitability supports its application where a completed structure, which has never been occupied by anyone, is placed for sale, even if the house was originally constructed for the builder's personal use and was the only house constructed by him in the state).

63. *Woodward*, 141 Ariz. at 516, 687 P.2d at 1271; *Richards v. Powercraft Homes, Inc.*, 139 Ariz. 242, 245, 678 P.2d 427, 430 (1984).

64. *Woodward*, 141 Ariz. at 516, 687 P.2d at 1271; *Richards*, 139 Ariz. at 245, 678 P.2d at 430; *Columbia Western Corp. v. Vela*, 122 Ariz. 28, 33, 592 P.2d 1299 (App. Div. 1 1979).

65. *Nastri*, 142 Ariz. at 441-42, 690 P.2d at 160–61. *See also, Sheibels v. Estes Homes*, 161 Ariz. 403, 404, 778 P.2d 1299, 1300 (App. 1989) (builder's implied warranty of habitability and workmanship extends to subsequent purchasers).

66. *Richards*, 139 Ariz. at 245, 678 P.2d at 430.

67. *Id.*

68. *Richards*, 139 Ariz. at 246, 678 P.2d at 431.

69. *See, Hershey v. Rich Rosen Const. Co.*, 169 Ariz. 110, 114, 817 P.2d 55, 59 (App. 1991) (reasonable inspection of residence need not include inspection by expert or professional home inspection service to scrutinize house for internal defects prior to purchase, but implied warranty will be voided for lack of reasonable inspection only if defect could have been discovered during inspection made by an average purchaser).

70. *S. Develop. Corp. v. Pima Capital Mgnt. Co.*, 201 Ariz. 10, 17, 31 P.3d 123, 130 (App. 2001).

71. *Hershey*, 169 Ariz. at 116, 817 P.2d at 61.

72. Mike Bischoff, "Theories of Toxic Mold Liability Facing Arizona Homebuilders," *Ariz. St. L. J.* 34:681, 700 (2002).

73. *Dillig v. Fisher*, 142 Ariz. 47, 51, 688 P.2d 693, 697 (App. 1984).

74. *Nastri v. Wood Bros. Homes, Inc.*, 142 Ariz. 439, 445, 690 P.2d 158, 164 (App. 1984). *See also, Arrow Leasing Corp. v. Cummins Arizona Diesel, Inc.*, 136 Ariz. 444, 666 P.2d 544 (App. 1983) (holding that recovery for property damage caused to a home by latent structural defects was not actionable on a product liability theory).

75. *Menendez v. Paddock Pool Const. Co.*, 172 Ariz. 258, 267, 836 P.2d 968, 977 (App. 1991) (swimming pool was not a "product" because swimming pool

had not been shown to be standard model constructed, assembled, or manufactured by mass-production process).

76. *See, Hall v. Lovell Regency Homes*, 121 Md. App. 1, 708 A.2d 344 (App. 1998) (damages for mental anguish caused by living in a less than defect-free house were subsumed in the proper measure of damages for a property damage claim).

77. *See, e.g., Dixon v. City of Phoenix*, 173 Ariz. 612, 618–19, 845 P.2d 1107, 1113–14 (App. 1992).

78. *See, e.g., DeStories v. City of Phoenix*, 154 Ariz. 604, 606, 744 P.2d 705, 707 (App. 1987) (a subcontractor's employees could not recover for mental anguish upon learning of their exposure to asbestos absent physical injury, illness, or other physical consequences).

79. *See, Erlich v. Menezes*, 87 Cal. Rptr.2d 886, 981 P.2d 978 (1999); Hall, 121 Md. App. 1, 708 A.2d 344.

80. *Fairway Builders, Inc. v. Malouf Towers Rental Co., Inc.*, 124 Ariz. 242, 253, 603 P.2d 513, 524 (App. 1979) (the measure of damages for breach of a construction contract is the cost of repair. Only where such an award would result in economic waste is the proper measure of damages the difference between the value of the building as designed and the building as constructed).

81. *See, e.g., Max of Switzerland, Inc. v. Allright Corp. of Delaware*, 187 Ariz. 496, 499, 930 P.2d 1010, 1013 (App. 1997) ("In Arizona, property damage claims for a negligently damaged vehicle include compensation for the cost of repair, residual diminution in fair market value, and loss of use").

82. *State v. Brockell*, 187 Ariz. 226, 228, 928 P.2d 650 (App. 1996); *Blanton & Co. v. Transamerica Title Ins. Co.*, 24 Ariz.App. 185, 188, 536 P.2d 1077, 1080 (1975); *A.I.D. Ins. Services v. Riley*, 25 Ariz.App. 132, 135-136, 541 P.2d 595, 598-599 (1975).

83. *Walter v. Simmons*, 169 Ariz. 229, 236, 818 P.2d 214, 221 (App. 1991) (Arizona law requires that damages be awarded only when the plaintiff has met its burden of proving the amount of damages with reasonable certainty).

84. *Allison v. Fire Ins. Exchange*, 98 S.W.3d 227 (Tex. App. 2002).

85. *Centex-Rooney Const. Co., Inc. v. Martin County*, 706 So.2d 20 (Fla. App. 1997).

86. *Allison*, 98 S.W.3d 227.

87. *Greene v. Plano, I.S.D.*, F.Supp. 2d 615 (E.D. Tex. 2002).

88. *RCDI Const., Inc. v. Space/Architecture Planning & Interiors, P.A.*, 29 Fed.Appx. 120 (4th Cir. 2002); *see also*, *Allison*, 98 S.W.3d 227.

89. *Allison*, 98 S.W.3d 227.

90. *Amica Mut. Ins. Co. v. Henderson*, 2003 WL 21196261 (N.D.Ill. 2003).

91. *Id.*

92. *Reedy v. Webb*, 2002 WL 3168023 (Tex. App. 2002).

93. California's "Toxic Mold Protection Act" was enacted in September 2001 by SB 732. All of California's regulations involving toxic mold can be found at Cal. Health & Safety § 26100 et al. (2003).

94. S.B. 1432, 45th Legislature, 2d Sess. (Ariz. 2002).

95. *Id.*

96. *Id.*

97. H.B. 2620, 45th Legislature, 2nd Sess. (Ariz. 2002).

98. A.R.S. § 12-1361(4).

99. A.R.S. § 12-1363(A).

100. A.R.S. § 12-1362(A).

101. A.R.S. § 12-1363(B).

102. *Id.*

103. *Id.*

104. *Id.*

105. A.R.S. § 12-1363(C).

106. *Id.*

107. *Id.*

108. A.R.S. § 12-1363(D).

109. A.R.S. § 12-1363(E).

110. *Id.*

111. *Id.*

112. *Id.*

113. A.R.S. § 12-1363(F).

114. A.R.S. § 12-1363(I).

115. A.R.S. § 12-1363(J).

116. A.R.S. § 12-1364.

117. A.R.S. § 12-1366.

118. *Cooper v. American Family Mut. Ins. Co.*, 184 F.Supp.2d 960 (D. Ariz. 2002).

119. *Id.* at 961

120. *Id.* The District Court specifically held that (1) the mold exclusion in the homeowners' insurance policy applied even if the covered water event was the efficient proximate cause of the mold damage; (2) damages caused by mycotoxins did not fall within the resulting loss provision; and (3) the supplementary coverage provision covering "pollutant" clean-up and removal did not modify the policy's mold exclusion.

121. *Id.* at 962.

122. *Id.* (emphasis added).

123. *Liristis v. American Family Mut. Ins. Co.*, 204 Ariz. 140, 61 P.3d 22 (App. 2002).

124. *Liristis*, 204 Ariz. at 142, 61 P.3d at 24.

125. *Liristis*, 204 Ariz. at 141, 61 P.3d at 23.

126. *Liristis*, 204 Ariz. at 144, 61 P.3d at 26.

127. *Id.*

128. *Liristis*, 204 Ariz. at 145, 61 P.3d at 27.

129. *Frye v. United States*, 293 F. 1013, 1014 (D.C. Cir. 1923).

130. *Logerquist v. McVey*, 196 Ariz. 470, 480, 1 P.3d 113, 123 (2000). In *Logerquist*, the expert testified that the patient had repressed memories of childhood sexual abuse for purposes of a medical malpractice action against a pediatrician. *Logerquist*, 196 Ariz. at 475–76, 1 P.3d at 118-19.

131. *Logerquist*, 196 Ariz. at 481, 1 P.3d at 124.

132. *State ex rel. Romley v. Fields*, 201 Ariz. 321, 325, 35 P.3d 82, 86 (App. 2001).

133. *Id.*

134. *Id.*

135. *See, e.g., id.* ("In Arizona, *Frye* has usually been applied in cases involving the results of physical scientific tests. *See, e.g., Valdez*, 91 Ariz. at 277–79, 371 P.2d at 896–98 (polygraph evidence); *Scales v. City Court*, 122 Ariz. 231, 234–35, 594 P.2d 97, 100-01 (1979) (breathalyzer); *State v. Gortarez*, 141 Ariz. 254, 262-66, 686 P.2d 1224, 1232–36 (1984) ("voiceprint" evidence); *State v. Superior Court (Blake)*, 149 Ariz. 269, 277–80, 718 P.2d 171, 179–82 (1986) (horizontal gaze nystagmus testing); *State v. Beaty*, 158 Ariz. 232, 241–42, 762 P.2d 519, 528–29 (1988) (phosphoglucomutase (PGM) blood grouping); *State v. Velasco*, 165 Ariz. 480, 486–87, 799 P.2d 821, 827–28 (1990) (silica gel blood alcohol test); *State v. Bible*, 175 Ariz. 549, 576-82, 858 P.2d 1152, 1179–85 (1993) (DNA evidence). However, *Frye* has also been applied to determine the basic reliability of techniques not necessarily involving "hard" science. *See State v. Mena*, 128 Ariz. 226, 231–32, 624 P.2d 1274, 1279-80 (1981) (hypnotically refreshed testimony). In a variety of other situations, however, *Frye* has been found inapplicable. *See, e.g., State v. Roscoe*, 145 Ariz. 212, 219, 700 P.2d 1312, 1319 (1984) (dog tracking); *Baroldy v. Ortho Pharm. Corp.*, 157 Ariz. 574, 581–83, 760 P.2d 574, 581–83 (App. 1988) (scientific hypothesis of causation); *State v. Richards*, 166 Ariz. 576, 577–79, 804 P.2d 109, 110–12 (App. 1990) (bite mark comparison); *State v. Varela*, 178 Ariz. 319, 325-26, 873 P.2d 657, 663-64 (App. 1993) (general characteristics of child sexual abuse victims).

136. *Logerquist*, 196 Ariz. at 482, 1 P.3d at 125.

137. *Lofgren v. Motorola*, 1998 WL 299925 (Ariz.Super. 1998), at 11 *citing Whiting v. Boston Edison Co.*, 891 F.Supp. 12, 24 (D.Mass.1995).

138. American College of Occupational and Environmental Medicine, "Adverse Human Health Effects Associated with Molds in the Indoor Environment," (2002) available at http://www.acoem.org/position/statements.asp?CATA_ID=52. *citing* R.S. Tobin et al., "Significance of Fungi in Indoor Air: Report of a Working Group," *Can. J. Public Health* 78:S1–S32 (1987); D. Menzies and J. Bourbeau, "Building-Related Illnesses," *N. Engl. J. Med.* 337:1524–31 (1997); F. Fung, R. Clark and S. Williams, "Stachybotrys: A Mycotoxin-Producing Fungus of Increasing Toxicologic Importance," *J. Toxicol. Clin. Toxicol.* 36: 79–86 (1998); C.A. Robbins et al., "Health Effects of Mycotoxins in Indoor Air: A Critical Review," *Appl. Occup. Environ. Hyg.* 15: 773–84 (2000); D.L. Sudakin, "Stachybotrys Chartarum: Current Knowledge of Its Role in Disease," *Med. Gen. Med.* E11 (2000); E.H. Page and D.B. Trout. "The Role of Stachybotrys Mycotoxins in Buildings Related Illness," *Am. Ind. Hyg. Assoc. J.* 62:644–8 (2001); Al Terr, "Stachybotrys: Relevance to Human Disease," *Ann. Allergy Asthma Immunol.* 87:57–63 (2001); H.A. Burge, "Fungi: Toxic Killers or Unavoidable Nuisances?" *Ann. Allergy Asthma Immunol.* 87:52–56 (2001).

Chapter 4

Mold in Delaware

Douglas F. Schleicher, Esq., and Bryan D. Weiss, Esq.

Synopsis

4.1 Introduction

Delaware courts and regulatory authorities have taken a relatively proactive role in addressing the perceived public health threats posed by mold exposure. For example, an entire apartment complex was condemned by the Town of Elsmere, Delaware causing the eviction of over 600 residents. The Delaware Supreme court has also upheld a combined award of over $1 million dollars against a Delaware landlord for mold-related personal injury claims. *See New Haverford Partnership v. Stroot et al.*, 772 A.2d 792 (Del. Spr. Ct. 2001) (hereinafter "*Stroot*"). This chapter will review these and other matters.

4.2 State and Local Enforcement Authority

To date, Delaware has not enacted legislation specifically addressing mold. In addition, the Delaware Department of Natural Resources and Environmental Control (DDEC) has not promulgated any specific regulations governing permissible mold exposure or remediation requirements. Notwithstanding the forgoing, there are several different governmental bodies which have enforcement authority relating to the presence of mold.

A. Delaware State Housing Authority

Pursuant to the Delaware State Housing Code (Code), the Housing Authority has broad authority to respond to "unsanitary" conditions existing in residential structures. As part of this broad power grant, the legislature has provided the Housing Authority with the power to examine and survey property in the state with respect to unsanitary conditions. If such conditions are identified, the Housing Authority can then order the alleged condition to be abated or removed. If the landowner refuses, the Housing Authority has the ability to abate the condition itself and file a tax lien on the property for its response costs. The landowner may be found guilty of a misdemeanor punishable by a maximum penalty of $1,000 and thirty days in jail. The Housing Authority also has the statutory authority to declare the premises to be "unfit for human habitation" and instigate condemnation and/or eviction proceedings. The Code specifically permits counties and municipalities to enforce its provisions.

B. Local cities and municipalities

In addition to providing a broad grant of authority to the Delaware State Housing Authority, the Code or variations thereof have been adopted by county, town and other municipal governments. Most of these local enactments provide local code officers with entry, inspection and abatement powers, including condemnation authority.

There are a number of examples of mold-related enforcement actions at the local level. For example, in the case of the Town of Elsmere, the purported authority for this action was based in the state Code and the local housing code. The town enforcement officer made a formal determination that mold on the premises rendered it unfit for human occupancy.

4.3 Private Rights of Action

A. Contractual claims for economic loss

The specific terms of a contract will govern any contract claim. Provisions relating to safety or cleanliness may apply to mold-related claims. Absent a provision pertaining to mold or sanitation, certain types of claims can be expected.

1. Implied warranty of habitability (residential)

Unlike many other American jurisdictions, Delaware has not historically recognized an implied warranty of habitability in real estate leases. *See Leech v. Husbands*, 152 A. 729 (Del. Spr. Ct. 1930); *see also Brandt v. Yeager*, 199 A.2d 768 (Del. Spr. Ct. 1964). Therefore, absent express contract language to the contrary, there is no implied contractual duty to provide safe and sanitary premises. Possibly to counteract this perceived problem, the legislature enacted the Delaware Landlord Tenant Code (LTC) in 1972. The LTC comprehensively governs the relationship of landlords and tenants. The LTC explicitly provides a private right of action to either party injured by a violation of a lease agreement or the LTC. Del. Code Ann. Tit. 25 §5117 (2002).

Relevant portions of the LTC require that landlords provide a rental unit that does not endanger the health, welfare or safety of the tenants or occupants. Landlords also are required to keep all common areas in a clean and sanitary condition. If a landlord fails to remedy violations of these provisions, the LTC provides the following potential remedies to the tenant:

- ability to terminate the lease;
- ability to remedy the problem and deduct relevant costs from outstanding rent; and
- keep two-thirds of rent accruing during the time of the violation (available only in limited cases).

Del. Code Ann. Tit. 25 §§ 5306-5308 (2002).

To date, there have not been any published cases involving mold in which a plaintiff has based its claim on the LTC's private right of action. There are several open issues regarding the remedy provisions of the LTC which have yet to be litigated. In *Brown v. Robyn Realty Co.*, 367 A.2d 183 (Del. Super. Ct. 1976), the Delaware Superior Court construed a since-repealed version of the LTC and held that: 1) the LTC provides a statutory right of action for general "damages", not limited to the remedies listed in the statute; and 2) the duties found in the LTC are "engrafted" onto all residential leases and can be the basis of a residential tenant's claim for breach of contract. In addition, the court noted that the rent abatement provisions might apply retroactively. *Id* at 190.

It is unclear whether the specific remedies in the current version of the LTC are exclusive. Landlords may be able to argue successfully that the remedies are exclusive because the legislature repealed the private right of action for general "damages" following the *Brown* decision. This is important because the new LTC requires that tenants prove actual or written notice of the unsanitary condition.

2. Warranty of habitability (commercial leases)

Commercial leases are normally assumed by courts to be the result of arms-length negotiations between parties with roughly equal bargaining power. Accordingly, many jurisdictions are unwilling to read a warranty of habitability into commercial leases. However, in one unreported decision involving a commercial lessee, a Delaware trial court noted that "[a] tenant has a right to claim breach of implied warranty of habitability as a defense or counterclaim." *See Groves v. E&H, Inc.*, 1981 WL 376978 (Del. Super. 1981) citing *Pugh v. Holmes*, 384 A.2d 1234 (Pa. Spr. Ct.) (Pennsylvania case establishing implied warranty of habitability in residential leases).

The original 1972 version of the LTC covered both commercial and residential leases. *See Hand v. Davis*, 1990 WL 96583 (Del. Super. Ct. 1990). However, during the 1996 revisions, the legislature explicitly exempted commercial leases from its coverage.

In light of the foregoing, commercial tenants should seek affirmative representations, warranties and covenants from the landlord regarding the sanitary condition of their space and other common areas of the building. In addition, they should consider identifying in the lease specific remedies that would be available in the case of a breach.

3. Implied warranty of construction in a skillful and workmanlike manner

Under Delaware law, construction contractors are deemed to provide an implied warranty that construction was performed in a reasonably workmanlike manner and that the finished product is fit for the purpose intended. *Bougourd v. Village Gardens Homes, Inc.*, 2002 WL 32072790 (Del. Com. Pl. 2002); *see also Smith v. Berwin Builders, Inc.*, 287 A.2d 693 (Del. Super. 1972). This warranty is based in contract law and results in regular contract damages in the case of a breach.

B. Torts

Tort claims can arise in a number of different contexts, such as landlord/tenant and premises liability matters. Entities such as design professionals (e.g., architects and engineers) and contractors may be targeted in property damage suits based on allegedly defective construction or design of buildings or HVAC systems.

1. Negligence (premises liability)

In *Stroot, supra*, two separate plaintiffs, Elizabeth Stroot and Joletta Watson, sought redress for bodily injury that they suffered as tenants of the defendant's apartment complex. Ms. Stroot was a thirty-three-year-old graduate student with a prior history of asthma and allergies that required hospital visits and previous medical attention. In May, 1994, she was hospitalized because she could not breath. The hospitalization coincided with the collapse of her bathroom ceiling, which exposed growing mold of various kinds. Ms. Watson had recurring mold in her bathroom. Watson testified that she experienced fatigue, sinus problems, headaches, chest pains and body aches which dissipated roughly six months after she moved from defendant's property.

The plaintiffs pursued three different theories of tort liability: negligence due to defendant's failure to maintain a safe and sanitary condition in the apartments; negligence *per se*, for violating the New Castle County Code, and negligence *per se* for violating the LTC. The jury verdict form indicated that the jury found for the plaintiffs on all three of their theories.

On appeal, the defendant argued that plaintiffs' ordinary negligence claim was cut off by Delaware's enactment of the LTC. The court disagreed noting that there was nothing in the LTC that suggested the legislature intended to cut off a plaintiff's common law tort claims. The court also noted that the plaintiffs could look to the LTC as a basis for establishing that the landlord owed a duty of care to its tenants to provide a safe and sanitary rental unit. The defendant also argued that the opinions of the plaintiffs' four different medical experts were deficient and should not have been presented to the jury, due to faulty protocols, insufficiency of corroborating data and inadequate rebuttal of the defendant's alternative explanations of the injuries. The Supreme Court held that the trial court has flexibility to evaluate the applicable *Daubert* factors under the circumstances presented in each case, and that the trial court had not abused its discretion.

2. Professional negligence

Professional negligence claims may arise against architects or engineers for personal injuries related to the alleged improper design or supervision relating to a building, its HVAC system, the building envelope or other features. *See Bonnette, et al. v. Tunica-Biloxi Indians*, No. 02-919, 2003 WL 21229282 (La. Ct. App. May, 28, 2003) (several groups of plaintiffs filed suit against construction defendants, including an architect and an engineering firm, for personal injuries from alleged mold exposure at a casino and its related child care facility). There are no reported Delaware decisions discussing the merits of a claim for professional liability for personal injury or property damage from mold. However, the Supreme Court of Delaware has generally recognized that architects and engineers have a duty to provide design services in accordance with the ordinary skill of their profession. *Seiler v. Levitz Furniture Company of Eastern Region, Inc.*, 367 A.2d 999 (Del. Spr. Ct. 1976). This duty extends to unrelated third parties and is not dependant on privity of contract. *Seiler* at 1007 (tenant of building was able to maintain suit against architect in negligence after suffering property damage from two floods).

3. Strict products liability

In addition to targeting the entities that own, manage, design or construct buildings, tort plaintiffs have also targeted manufacturers of exterior cladding systems using strict product liability theories. *See Gropper, et al. v. STO Corporation, et al.*, 552 S.E.2d 118 (Ga. App. 2001) (homeowners filed suit against builder, stucco subcontractor and stucco manufacturer asserting claims of strict products liability for mold-related personal injuries and property damage due to use of synthetic stucco product on the exterior of their home). However, Delaware does not recognize a common law right of action for strict products liability based on the sale of defective goods under Section 402A of the Restatement (Second) of Torts (1965). *See Cline v. Prowler Industries of Maryland, Inc.*, 418 A.2d 968 (Del. 1980) (Supreme Court of Delaware held that the Uniform Commercial Code pre-empted strict products liability theory in Delaware).

4. Negligent misrepresentation (professional)

Mold litigation often involves finger pointing between contractors and professionals with respect to responsibility for the specific conditions leading to mold growth. Contractors may allege defects in the designs of the professionals associated with the project. This can be done via claims based on the Restatement (Second) of Torts §552 arguing that the design professional's plans and specifications incorporated "negligent misrepresentations." In order to state a claim for negligent misrepresentation a contractor must show misrepresentation of a material fact; that the defendant either knows it is a misrepresentation, makes it without knowledge of its truth or falsity, or makes the representation under circumstances in which it ought to know of its falsity; that the defendant must intend the representation to induce another to act on the misrepresentation; and that the contractor suffered an injury acting in justifiable reliance on the misrepresentation. *See Guardian Construction Co. v. Tetra Tech Richardson, Inc.*, 583 A.2d 1378 (Del. Super. 1990); *Counsel of Dorset Condominium Apartments v. Dorset Apartments*, 1992 WL 240444 (Del. Super 1992).

At this time, there are no reported cases in Delaware involving a contractor's claims of negligent misrepresentation against a design professional related to mold. However, a contractor will be provided with an opportunity to prove these claims even if they are based on strictly economic loss. *See Guardian, supra* (economic loss doctrine not applicable to negligent misrepresentation claims based on §552 of the Restatement [Second] of Torts).

5. Implied warranty of fitness for intended purpose (professionals)

Although a contractor's action for negligent misrepresentation against an architect is not precluded by the economic loss doctrine under Delaware law, design professionals may be insulated from claims that they impliedly warrant that their plans and specifications will result in a structure reasonably fit for its intended purpose. *See, e.g., Grotto Pizza, Inc. v. Endecon, Inc.*, 1997 WL 366904 (Del. Super. 1997). Therefore, in the absence of any express warranties in contract, owners may be limited to suing contractors and indirectly accessing the professional by way of the contractor's own third-party negligent misrepresentation claim for errors or omissions.

6. Negligent workmanship/construction

Contractors and construction managers are likely to be targeted directly by owners in property damage claims related to mold and moisture conditions. S*ee Centex-Rooney, supra* (Florida appellate court upheld $8.8 million dollar breach of contract award against construction manager for failing to properly supervise construction). To date, there are no reported Delaware opinions discussing claims that can be brought against contractors arising from mold. However, Delaware courts generally recognize several different theories of redress against contractors for construction defects, including an action for "negligent workmanship." *See Taylor, supra* at 4 (contractor has a common law duty to perform its work in a reasonable manner); s*ee also Oliver B. Cannon & Sons, Inc., v. Dorr-Oliver Inc.*, 312 A.2d 322, 325 (Del. Super. 1973).

7. Fraudulent misrepresentation (sale of property)

Fraudulent misrepresentation is actionable if the following is shown: (1) a representation is made; (2) with knowledge of its falsity or recklessness as to whether it is true or false; (3) with the intent to induce the plaintiff to act or to refrain from acting; (4) plaintiff acts or fails to act in justifiable reliance on the representation; and (5) the resulting injury was proximately caused by the reliance. *Stephenson v. Capano*, 462 A.2d 1069, 1074 (Del. 1999). The tort of fraudulent concealment has the same elements as fraudulent misrepresentation, except that with concealment the party intentionally conceals, rather than affirmatively misrepresents, the material fact. *Lock v. Schreppler*, 426 A.2d 856, 860 (Del. Super. 1981). While concealment may be fraudulent, mere silence is not sufficient in the absence of a duty to speak. *The Marcom Group v. Redgo Properties, Inc.*, 1998 WL 278368 (Del. Super. 1998). Normally, fraudulent concealment is not actionable unless there is active conduct.

A real estate broker, in addition to a seller, can be liable for the torts of fraudulent non-disclosure. See *Lock, supra*. In *Lock*, the court refused to dispose of the plaintiff's claim and held that a broker had an affirmative duty to disclose all relevant information regarding a known termite infestation once the broker had raised the subject in discussions with the purchasers. *Lock, supra* at 861.

8. Negligent misrepresentation (sale of property)

To successfully claim negligent misrepresentation, the plaintiff must prove: (1) a pecuniary duty to provide accurate information; (2) the provision of false information; (3) the failure to exercise reasonable care in obtaining or communicating the information; and (4) pecuniary loss to a party acting in justifiable reliance on the misrepresentation. *Durnam v. Manallo, 2004 WL 326924 (Del. Com. Pl. 2004).* Unlike intentional misrepresentation, the speaker need not know his or her words are untrue, but must have failed to make a reasonable investigation of their truth.

9. Innocent misrepresentation (sale of property)

An innocent misrepresentation is actionable if it is material to and the basis of a transaction, and the purchaser is justified in relying on the misrepresentation. *Norton v. Poplos*, 443 A.2d 1 (Del. 1982). However, the remedy for an innocent misrepresentation is rescission, not monetary damages for tort recovery. *Id.*

Chapter 5

An Overview of "Toxic" Mold Litigation in Florida

Leslie O'Neal-Coble

Synopsis

5.1 Introduction

Florida's warm, humid climate provides perfect conditions for mold to flourish. Consequently, it is no surprise that mold-related litigation flourishes also. Florida has the distinction of having some of the earliest mold-related cases to go through trial and appeal. Florida courts were among the first to grapple with *Frye*[1] challenges to expert testimony, certification of class actions by tenants,[2] and application of workers' compensation immunity[3] in mold-related litigation.

There are three main aspects to mold-related litigation: (1) personal injury claims; (2) property damage and building defect claims; (3) insurance coverage issues. In addition, as noted above, there are numerous procedural and evidentiary issues involved in these cases. This chapter will discuss the current status of the law in Florida on these issues.

5.2 Personal Injury Claims

There are no appellate decisions in Florida specifically regarding the theories that may be used to assert personal injury claims for mold exposure.

Reported trial court opinions have allowed plaintiffs to pursue various legal theories, including negligence and intentional tort[4] as well as breach of implied warranty of merchantability and fitness for a particular purpose and for violation of the Florida Building Code.[5]

Unlike some states, Florida has not had any substantial jury verdicts for personal injury claims related to mold exposure yet. This may be in part because Florida trial courts routinely require pre-trial mediation[6] in all civil cases and these mediations are remarkably successful in settling cases.

5.3 Workers' Compensation Claims

Florida's Workers Compensation Act has been amended to limit employers' liability for mold-related injuries. Section 440.02(1) provides that:

> An injury or disease caused by exposure to a toxic substance, including but not limited to fungus or mold, is not an injury by accident arising out of employment unless there is clear and convincing evidence establishing that exposure to the specific substance involved, at the levels in which the employee was exposed, can cause the injury or disease sustained by the employee.

This is a very difficult standard.

In Florida, workers' compensation is an employee's exclusive remedy against the employer for a workplace injury.[7] The workers' compensation exclusivity includes

> vicarious liability of such employer to any third-party tortfeasor and to the employee, the legal representative thereof, husband or wife, parents, dependents, next of kin, and anyone otherwise entitled to recover damages from such employer at law or in admiralty on account of such injury or death…

There are some exceptions to the employer's immunity defense, such as: (1) the employer's failure to pay for compensation benefits; (2) the employer commits an intentional tort against the employee, causing the employee's injury; (3) the employer intentionally exposes the employee to a condition that was substantially certain to cause injury, but misrepresenting or concealing the danger from the employee.[8] The employer's immunity also extends to fellow employees, unless they act, "with willful and wanton disregard or unprovoked physical aggression or with gross negligence when such acts result in injury or death or such acts proximately cause such injury or death."[9] The immunity does not apply to employees of the same employer, "when each is operating in the furtherance of the employer's business but they are assigned primarily to unrelated works within private or public employment."

In *McClanahan v. State*, [10] several current and former employees of the Florida Department of Juvenile Justice (DJJ) alleged they became ill because of unhealthy and unsafe air quality in the building where they worked. They sued the DJJ and the Florida Department of Management Services (DMS) (the state agency responsible for public building construction) and others, alleging that their negligence in the design and construction of the building caused the plaintiffs' injuries. The complaint sought damages for respiratory injuries suffered by the employees and for the wrongful death of one employee.

The trial court granted summary judgment in favor of the DJJ and the DMS and the plaintiffs appealed. The First District Court of Appeals affirmed the summary judgment, finding that the plaintiffs' allegations did not support a conclusion that the state agencies' acts or omissions were "substantially certain to result in injury or death." At most, the evidence "might support a conclusion that the agencies negligently exposed the employees to increased risk," by failing to take steps to improve the air quality in the building and that the agencies were "less than candid" with their employees. Further, the state's immunity included DMS, even though it did not employ the plaintiffs.

The Florida Supreme Court reaffirmed the intentional tort exception to workers' compensation immunity in *Turner v. PCR, Inc.* [11] holding that an employer may be held liable in tort if it has engaged in conduct that was intended to or was substantially certain to result in an employee's injury or death. The *Turner* court also held that when applying the "substantial certainty of injury" standard, the employer's conduct must be evaluated under an objective test rather than a subjective one. The employer may be held liable if "it should have known ... that the conduct complained of was "substantially certain to result in injury or death."[12] Proving the intentional tort exception requires a showing that the employer engaged in conduct that is at least worse than "gross negligence."[13] This is a difficult standard and will rarely apply in mold exposure cases.

5.4 Property Damage and Building Defect Claims

The leading Florida case on building defect claims related to mold growth is *Centex-Rooney Construction Co. v. Martin County.*[14] The county sued Centex-Rooney, the construction manager on the county's courthouse project, for breach of contract because of significant construction defects in the buildings which led to water intrusion, high humidity and mold growth.

Shortly after occupying the buildings in December 1988, employees began complaining about high humidity, water leaks and mold growth. The county made demands on Centex-Rooney to remedy the problems under the terms of its warranty, but the problems continued. The county also received health complaints from building occupants and visitors. Within four years after initial occupancy, 25 percent of the occupants had evacuated the buildings for health reasons. By December, 1992, the courthouse advisory committee, comprised of judges and other officials who occupied the buildings, voted to evacuate because they did not believe they could, "in good faith and good conscience, ask the county's employees, the litigants, the attorneys and the general public to enter these buildings under the existing conditions."[15] Consequently, all of the judges, clerks, and related employees were moved to temporary facilities while the mold and related problems were repaired.

The county hired a remediation firm to remove visible mold growth on exterior walls in specified areas of the buildings. However, the remediation revealed more mold and water damage than originally anticipated, which resulted in extensive demolition of the buildings. In addition, remediation revealed numerous other hidden and dangerous electrical and structural defects in the buildings which had to be repaired.[16] All of this expanded the scope of the remediation and repair work and increased the cost for redesign, reconstruction and relocation. The repairs took over three years to complete and were still underway at the time of trial.

The county sued the construction manager, its performance bond sureties, the architect, the mechanical engineer, and the concrete and masonry construction company for breach of contract and negligence. Prior to trial, the county settled with all the defendants except Centex-Rooney and its sureties.

After a five and one-half week trial, the jury returned a verdict for the county in the amount of $11,550,000. This amount was reduced by $2,750,000, the amount of the county's prior settlements. The court awarded on prejudgment interest $8,800,000 from December 12, 1988, the date of substantial completion, through June 18, 1996, resulting in a final judgment of over $14 million.[17]

One of Centex-Rooney's primary issues on appeal was that the county had failed to prove that the construction defects caused an actual health hazard justifying evacuation of the buildings. Rejecting this argument, the appellate court stated

> Centex's claim that the County was required to prove that the construction defects caused an actual health hazard misses the mark . . . We think it is important to observe that Centex has never contested the fact that there were serious and substantial defects in the buildings. The reason for the evacuation and whether that reason was supported by the evidence was immaterial to the elements that the County was required to prove in its contract action.[18]

The county took the position that it was not required to prove that the conditions in the buildings actually caused anyone to become ill, but rather that the conditions presented a potential health threat that required the county to take reasonable measures to remove the mold and to repair the water intrusion and humidity problems.

On appeal, Centex-Rooney and the sureties also asserted that the trial court erred in allowing the testimony of two county experts, Dr. Phil Morey and Dr. Michael Hodgson, regarding scientific evidence of the potential "health hazard stemming from the toxic molds found in the buildings."[19] The trial court conducted *Frye* hearings before allowing these experts to testify. The appellate court, conducting a *de novo* review of the evidence from the *Frye* hearings, held that the county "had met its burden of proving that the basic underlying principles of scientific evidence were sufficiently tested and accepted by the relevant scientific community."[20]

The appellate court also rejected Centex-Rooney's claim that the county failed to present sufficient evidence to support its damages claim. The court noted that in a breach of contract case, a measure of damages is the reasonable cost of performing construction and repairs in conformance with the original contract requirements, plus relocation and financing costs and additional engineering and architectural fees necessary to accomplish the reconstruction.[21]

The county had several witnesses testify about the actual costs incurred for demolition, redesign and reconstruction of the buildings, relocation and financing. The county introduced boxes of vouchers and checks showing the amounts paid for these costs. The county's damage expert reviewed reports of these expenses and analyzed them, removing all costs related to upgrades or questionable expenses. The trial court allowed the expert to testify from a summary report that recapitulated the county's total expenses and claimed damages. The expert then deflated the county's claimed damages to 1988 dollars and calculated a single present day value as of the date of the breach, which was stipulated to be December 12, 1988, the date of substantial completion. On appeal, the court dismissed Centex's challenges to this testimony, noting that:

> [u]ncertainty as to the amount of damages or difficulty in proving the exact amount will not prevent recovery where it is clear that substantial damages were suffered and there is a reasonable basis in the evidence for the amount awarded. Ultimately the degree of certainty simply requires that the mind of a prudent impartial person be satisfied with the damages.[22]

The court noted that:

> the County's unrefuted evidence of its actual costs to remediate, redesign and reconstruct the buildings as well as relocate and finance, reduced by the costs for upgrades and improvements, established a sufficient basis to determine damages as of the date of the breach.[23]

Defendants in property damage claims for mold growth often raise Florida's economic loss rule as a defense. The economic loss rule is a "court created doctrine which prohibits the extension of tort recovery for cases in which a product has damaged only itself and there is no personal injury or damage to 'other property,' and the losses or damages are economic in nature."[24] While the economic loss rule may prevent recovery under a negligence theory for damage to the structure, it will not preclude recovery for personal injuries or for property damage to personal property unconnected with the sale of the structure.[25]

In *Casa Clara Condominium Assoc., Inc. v. Charley Toppino and Sons, Inc.*[26] the Florida Supreme Court held that the economic loss rule applied to the purchase of houses[27] and condominiums and barred a negligence claim by a condominium association against a concrete supplier. The court rejected the association's argument that the defective concrete "damaged other property because the individual components and items of building material, not the homes themselves, are the products they purchased."[28] The court stated:

> The character of a loss determines the appropriate remedies and, to determine the character of a loss, one must look to the product purchased by the plaintiff, not the product sold by the defendant. [citation omitted] ... These homeowners bought finished products—dwellings—not the individual components of those dwellings. They bargained for the finished products, not their various components. The concrete became an integral part of the finished product and, thus did not injure 'other' property.[29]

However, the Florida Supreme Court retreated from its position regarding application of the economic loss rule in recent decisions.[30] In *Moransais v. Heathman*[31] the Florida Supreme Court held that homeowners could sue engineers individually in negligence, "despite the lack of a direct contract" between the parties and even though the homeowners had direct contract remedies against the seller and the engineering firm that employed the individual engineers.[32]

In *Comptech Int'l, Inc. v. Milam Commerce Park, Ltd.*,[33] decided four months after *Moransais*, the Florida Supreme Court held that the economic loss rule did not apply to a claim by a tenant against a landlord for negligence in hiring a contractor, violation of the Florida Building Code,[34] and return of illegally collected rent. The Florida Supreme Court held that the economic loss rule did not apply, at least in part, because the case involved a statutorily created cause of action.[35] A recent case from the Eleventh Circuit Court of Appeals has certified several questions regarding application of the economic loss rule to the Florida Supreme Court.[36] Until there is a definitive decision, the application of the economic loss rule remains a source of confusion for plaintiffs and defendants in mold litigation.

5.5 Insurance Issues

The issue of "property damage" to "other property" also arises regarding insurance coverage for defective construction that causes water intrusion and mold damage. In *Lassiter Construction Co. v. American States Insurance Co.*[37] the court held there was no insurance coverage for a complaint for defective construction which failed to allege personal injuries or damage to property other than buildings themselves.

The Florida Supreme Court has held that, "[A]s a matter of public policy, a commercial liability insurance contract, like Aetna's, does not cover claims for defective or deficient workmanship."[38] The purpose of comprehensive liability insurance coverage is to provide protection for personal injury or for property damage *caused by* the completed product, *but not for the replacement and repair of that product*.[39] To hold otherwise, would allow a contractor to receive an initial payment for its work, and also claim insurance coverage to repair and correct deficiencies in that work.[40]

5.6 Florida's Construction Defect Law

Under a new law that became effective May 27, 2003,[41] owners of dwellings in Florida must give contractors notice and an opportunity to cure construction defects before filing suit. This will include defects resulting in water intrusion, high humidity and mold growth. The law only applies to residential dwellings, and to certain kinds of defects. It establishes a rigid timeframe that both parties must follow to preserve their rights. In theory the law should reduce litigation by requiring the parties to try to resolve their differences before heading to court. In practice, it may not be that simple. The law was amended in 2004, the amendments were effective July 1, 2004.

The new law applies only to dwellings. "Dwellings" include single family homes, multifamily buildings, and condominiums, including common areas, but not commercial structures. "Claimants" required to give notice of defects under the law include original purchasers, subsequent purchasers, renters, and associations, but not contractors. "Contractors" entitled to receive notice of defects include general contractors, subcontractors, design professionals, developers, and anyone else "legally engaged in the business of designing, developing, constructing, manufacturing, selling, or remodeling dwellings." By specifying that contractors be "legally engaged" in business, the legislature apparently intended not to require claimants to give notice to unlicensed contractors.

The law covers defects in new construction and remodeling, and repairs. Moreover, only certain kinds of defects are covered. The claimant need only give notice of defects involving defective materials, violation of applicable building codes, design malpractice, or failure to employ workmanlike construction methods. The legislative staff report makes clear that defects not meeting the statutory definition are not subject to the notice provisions of the law. An example of a defect not subject to the notice provisions of the act might be an erroneous but workmanlike installation of galvanized downspouts when aluminum was specified. Finally, the law applies only to claims for damage or loss to the dwelling or to personal property, but not to personal injuries.

In a case that meets all of these statutory definitions, here is how the law works. The claimant must give any contractor believed responsible at least sixty days written notice of the claimed defects before filing suit against the contractor. The notice must be either hand delivered or sent by certified mail and must specify the defect or defects in reasonable detail. Serving the notice tolls the running of the statute of limitations. The law encourages the claimant to "endeavor to serve the notice of claim within fifteen days" after discovering the defect, but a failure to do so does not bar the claimant's right to sue. If the claimant sues before giving any notice, however, or without complying with the other terms of the law, the contractor can have the suit abated (i.e., suspended) until the claimant fully complies with the law.

Once the contractor receives notice of a claimed defect, the contractor has thirty days to conduct an inspection of the dwelling, although inspection is not mandatory. The inspection may include destructive testing by agreement with the claimant. Within ten days after receiving notice, the contractor may forward the notice to any subcontractor, supplier, or design professional who the contractor believes is responsible for the defects. These parties then have fifteen days to conduct their own inspection of the dwelling. In addition, within the same five business days, these parties must provide the contractor with a written report describing the results of their inspection and stating whether they accept responsibility for the defects and what they will do to repair them.

Within forty-five days after receiving the claimant's notice of defects, each contractor receiving notice must serve a written response on the claimant. The response must either

- offer to remedy the defect at no cost to the owner, including a detailed report on the contractor's inspection and findings, the work to be performed, and the timetable for completion of the work;
- offer to settle the claim for a monetary payment to be paid within a specified time; or
- state that the contractor disputes the claim and will neither remedy the defect nor settle the claim; or
- a written statement that a monetary payment, including insurance proceeds, if any, will be determined by the person's insurer within 30 days after notification to the insurer. The parties may agree to a partial settlement of the claim.

If the contractor disputes the claim, the claimant may sue without further notice. On the other hand, if the contractor's response offers to repair the defects or settle the claim, the response must include a statement advising the claimant that the contractor's offer will be deemed accepted if not rejected in writing within forty-five days after receipt of the offer. In order to properly reject the contractor's offer, the claimant must serve the contractor with written notice of rejection that includes the contractor's offer with the word "rejected" printed on the offer. After rejecting the offer, the claimant may sue without further notice.

If the claimant accepts the contractor's offer to repair the defects or settle the claim, and the contractor complies with the terms of the offer, then the claimant is barred from suing on the claim. On the other hand, if the contractor does not meet the terms of its offer, either by not paying the settlement money or by not making the repairs promised within the agreed time or in the agreed manner, then the claimant may sue without further notice. The failure of any party to comply with the law is admissible in any subsequent suit.

The procedures of the act supercede any contrary provisions of an arbitration clause in the contract between the parties. Mailing a written notice of claim tolls the applicable statute of limitation for 90 or 120 days.

In theory, these procedures should lead to mutually beneficial out-of-court resolutions of dwelling defect claims. In practice, however, the law is likely to create problems for both claimants and contractors.

The revised law gives contractors and subcontractors thirty business days after receiving notice to conduct an inspection of the dwelling. The initial list of construction defects may be amended to include additional or new construction defects. The court shall allow actions to proceed as to alledged construction defects that were noticed and for which the claimant has complied with this act.

Similarly, the law places the strict requirement on the claimant to properly reject the contractor's offer to repair the defect or settle the claim within forty-five days, or the offer is deemed accepted. Even though the contractor's offer must advise the claimant of this burden, many unsophisticated claimants will likely miss the required deadline or fail to serve a statutorily adequate rejection, and thereby lose their right to sue. Furthermore, for all but the simplest of defects, the claimant may not be able, in fifteen days, to obtain the independent advice needed to evaluate the adequacy of the contractor's offer. If the contractor's offer is inadequate, and the claimant either accepts the offer or fails to timely and properly reject it, the claimant has no other recourse and will have to settle for whatever the contractor offered.

A written report setting forth the details of the contractor's inspection and findings. The revised law provides that both claimants and contractors or others receiving notice have a mutual duty to exchange all available discoverable evidence relating to the construction defects, including, but not limited to, expert reports, photographs, and other information. Parties who fail to provide such evidence shall be subject to sanctions as the court may impose for discovery violations. Expert reports exchanged may not be used in subsequent litigation unless the expert testifies as a witness or the report is relied upon by a testifying expert for the party providing the report.

These and perhaps other terms of the new law will likely generate some controversy in the courts. Moreover, all parties will almost certainly need legal counsel to assist them in navigating these new waters, a fact that cuts against the bill's purpose of avoiding legal entanglement. On the whole, however, this new act will have the salutary effect of protecting contractors from the expense of litigation and the risk of excessive damage awards where the contractor is willing to satisfy the owner through a monetary payment or by repairing defects at no cost to the owner.

5.7 Class Actions

A recent trial court case certified a class in a claim by tenants in an apartment building against the landlord alleging personal injury and property damage because of mold.[42] The trial court certified the class despite the landlord's argument that the issues—injury, causation, apportionment of liability, defenses and damages—cannot be tried on a classwide basis because individual issues predominate.[43] This is the only Florida case, of which the author is aware, certifying a class in a "mold" litigation matter. In light of the recent decision by the Florida Third District Court of Appeal reversing class certification in a lawsuit by smokers against a tobacco company, it would seem that there will be few classes that will be certified.[44] The recently enacted Class Action Fairness Act of 2005 may also affect class action litigation in mold cases.

5.8 Conclusion

As long as mold continues to grow in Florida, Florida courts will continue to deal with the complex issues involved in mold litigation. Both plaintiffs' and defense attorneys will have the opportunity to create new laws in this field for many years to come.

Endnotes

1. *Frye v. U.S.*, 293 F. 1013 (D.C. Cir 1923). Florida follows the *Frye* standard regarding admissibility of expert testimony rather than the standard in *Daubert v. Merrell Dow Pharmaceuticals, Inc.*, 509 U.S. 579, 592–593, 113 S.Ct. 2786 (1993). *See Centex-Rooney Construction Co. v. Martin County*, 706 So.2d 20 (Fla. App. 1997), *rev. dismissed*, 718 So.2d 1233 (Fla. 1998).

2. *Archstone-Smith Operating Trust v. Henriques*, Case No: 3D03-403, In the District Court of Appeal, State of Florida, Third District.

3. *McClanahan v. State*, 28 F.L.W. D2078a (Fla.App. 2003).

4. *Id.*

5. *Savage v. WCI Communities, Inc.*, 10 Fla. L. W. Supp. 621a (Case No: 02-20505 09, Circuit Court 17th Judicial Circuit in and for Broward County, Florida); order on motion to dismiss entered June 9, 2003). FLA. STAT. §553.84 provides a civil right of action to persons aggrieved by violations of the Florida Building Code.

6. *See* FLA. STAT. §44.102 (2003).

7. *See* FLA. STAT. §440.11(1)(b) (2003).

8. *Id.*

9. *Id.*

10. *McClanahan v. State*, 28 F.L.W. D2078a (Fla. App. 2003).

11. *Turner v. PCR, Inc.*, 754 So.2d 683 (Fla. 2000).

12. *Id.* at 688-89.

13. *Id.* at 687, footnote 4.

14. *Centex-Rooney Construction Co. v. Martin County*, 706 So.2d 20 (Fla. App. 1997), *rev. dismissed,* 718 So.2d 1233 (Fla. 1998).

15. *Id.* at 24.

16. These defects included the absence of necessary expansion joints in the EIFS system; an inadequate number and improper spacing of EIFS system fasteners; severe rust and water damage of metal studs, fasteners and other building components; the inadequate attachment of multi-story glass curtain walls; numerous cuts in the curtain wall frames; improper window glass and frame installation; drainage system defects; and over 1,200 potential electrical system defects. *Id.* at 25, footnote 2.

17. *Id.* at 25.

18. *Id.*

19. *Id.* at 26.

20. *Id.*

21. *Id.* [citations omitted].

22. *Id.* at 28, *citing Adams v. Dreyfus Interstate Dev. Corp.*, 352 So.2d 76, 78 (Fla. App. 1977).

23. *Id.*

24. *Comptech International, Inc. v. Milam Commerce Park, Ltd.* 753 1219, 1221 (Fla. 2000).

25. *See, e.g., Savage v. WCI Communities*, supra, note v.

26. *Casa Clara Condominium Assoc., Inc. v. Charley Toppino and Sons, Inc.*, 620 So.2d 1244 (Fla. 1993).

27. The *Casa Clara* court specifically disapproved the contrary holdings in *Latite Roofing Co. v. Urbanek*, 528 So.2d 1381 (Fla. App. 1988); *Drexel Properties, Inc. v. Bay Colony Club Condominium, Inc.* 406 So.2d 5151 (Fla. App. 1981), rev. den. 417 So.2d 328 (Fla. 1982) and *Adobe Building Centers, Inc. v. Reynolds*, 403 So.2d 1033 (Fla. App. 1980), *rev. dismissed,* 411 So.2d 380 (Fla. 1981).

28. *Id.* at 1247.

29. *Id.*

30. *See generally,* Charles R. Walker, Note, "*Moransais v. Heathman* and the Florida Economic Loss Rule: Attempting to Leash the Tort-Eating Monster," 52:769 *Fla. L. Rev.* (2000).

31. *Moransais v. Heathman,* 744 So.2d 973 (Fla. 1999).

32. *Id.* at 975, 984.

33. *Comptech Int'l, Inc. v. Milam Commerce Park, Ltd.*, 753 So.2d 1219 (Fla. 1999).

34. Section 553.84 FLA. STAT. provides:

> Statutory civil action.—Notwithstanding any other remedies available, any person or party, in an individual capacity or on behalf of a class of persons or parties, damaged as a result of a violation of this part of the Florida Building Code, has a cause of action in any court of competent jurisdiction against the person or party who committed the violation; however, if the person or party obtains the required building permits and any local government or public agency with authority to enforce the Florida Building Code approves the

plans, if the construction project passes all required inspections under the code, and if there is no personal injury or damage to property other than the property that is the subject of the permits, plans, and inspections, this section does not apply unless the person or party knew or should have known that the violation existed.

35. *Id.* at 1223.

36. *Indemnity Insurance Company of North America v. American Aviation, Inc.*, 2003 U.S. App. LEXIS 18367, 16 Fla. L. Weekly Fed. C 1057 (2003).

37. *Lassiter Construction Co. v. American States Insurance Co.*, 699 So.2d 768 (Fla. App. 1997).

38. *See LaMarche v. Shelby Mut. Ins. Co.*, 390 So. 2d 325 (Fla. 1980).

39. *Id.* at 326.

40. *Id.*

41. The new law has been assigned to Chapter 558.001-558.007 of the Florida Statutes. Text of the bill, designated CS/CS/SB 1286, may be viewed at www.flsenate.gov.

42. *Henriques v. Archstone-Smith Operating Trust*, *supra*, note ii.

43. Fla. R. Civ. P. 1.220 (b)(3); *Liggett Group, Inc. v. Engle*, No. 3D00-3400, slip op. at 6-24 (Fla. 3d DCA May, 21, 2003) (case is unsuitable for class-action treatment where liability, defenses and damages depended on resolution of individual issues).

44. *Liggett Group, Inc. v. Engle*, 853 So. 2d434 (Fla. 3d Oct. 2003), *review granted* 873 So. 2d 1222 (Fla. 2004).

Chapter 6

Mold Litigation in Illinois

Edward Eshoo, Jr.

Synopsis

6.1 Introduction

Illinois is no stranger to mold litigation. An ever increasing number of mold-related personal injury and property damage lawsuits are being filed in Illinois against property insurers, building product manufacturers, contractors, mold abatement companies, indoor air quality professionals, as well as owners and sellers of real property. Even Michael Jordan, arguably Illinois' most famous resident over the past fifteen years, has joined the parade of litigants.[1] This chapter will examine legal issues confronting mold litigants in Illinois.

6.2 First-Party Property Insurance Coverage

A first-party property insurance policy provides coverage for loss or damage to the insured's covered property caused by a covered cause of loss. There are no reported appellate decisions in Illinois concerning first-party property insurance coverage for mold damage. The reason is because insurers traditionally have covered mold damage if it is the result of a covered peril.[2]

Although there are no published appellate decisions concerning first-party property insurance coverage for mold damage, the Illinois Department of Insurance has issued a bulletin advising as follows:

> All personal and commercial property policies that provide coverage for the perils of fire or lightning (pursuant to Class 3 (a) of Section 4 of the Illinois Insurance Code) must cover mold, fungus, etc. that results from a covered fire or lightning loss. If there is water damage and mold following a covered fire or lightning loss, the total of all damages is subject to the full policy limit.[3]

The Department's position is based on the 165-line, 1943 New York Standard Fire Policy (SFP). Illinois has adopted the SFP by statute. All fire insurance policies written in the State of Illinois must conform to the requirements of the SFP.[4] As such, the SFP sets forth minimum fire insurance coverage upon which an insured can rely under any fire insurance policy issued in the State of Illinois. The SFP guarantees a minimum level of coverage that supersedes any attempt to limit coverage to less than the statutory minimum. A fire insurance policy may not provide less coverage than that set forth in the SFP. Stated differently, a fire insurance policy must provide coverage in conformity with or in excess of the SFP. The SFP thus prohibits the use in fire insurance policies of limitations, conditions, or exclusions not expressly authorized.[5]

The SFP provides coverage for all direct loss by fire and lightning.[6] The perils not covered include such things as loss caused by enemy attack, invasion, insurrection, rebellion, revolution, and civil war.[7] The SFP neither limits nor excludes coverage for mold damage. Thus, under the SFP, water damage and mold damage resulting from a covered fire or lightning loss is covered.

There have been a number of legal challenges outside of Illinois involving first-party property insurance coverage for mold damage. The issues in many of these cases involved (a) whether mold spores constitute physical loss or damage, as required by the insuring agreement for most first-party property insurance policies, (b) whether coverage for mold damage is barred by the exclusions for (1) loss caused by rust, corrosion, mold, wet or dry rot, (2) loss caused by the discharge, dispersal, seepage, migration, release, or escape of pollutants, which term is usually defined to include any solid, liquid, gaseous or thermal irritant or contaminant, and (3) whether coverage is afforded for water and mold damage which ensues from faulty, inadequate, or defective workmanship, construction, or maintenance. This author submits that, under Illinois law, mold spores constitute physical loss or damage. This author submits that, under Illinois law, neither the "pollution" exclusion nor the "mold" exclusion should be a bar to coverage for mold damage.[8] This author also submits that, under Illinois law, water and mold damage which ensues from faulty, inadequate, or defective workmanship, construction, or maintenance is covered.

In *Board of Education of Township High School District No. 211 v. International Insurance Co.*,[9] the applicable policies provided coverage for "all risks of physical loss or damage to all real or personal property of every kind or description wherever located occurring during the period of this insurance." The issues were whether the presence of friable asbestos-containing building materials and asbestos fibers constituted physical loss or damage to covered property and whether such physical loss or damage occurred during the period the policies were in effect. The appellate court concluded that asbestos fiber contamination in plaintiffs' schools constituted physical loss or damage. The appellate court also concluded that the insureds' claim for asbestos removal costs incurred during the policy period "triggered" coverage. In reaching this conclusion, the appellate court applied the "equitable continuous trigger." Under that trigger, asbestos property damage is deemed an ongoing, continuous process triggering every policy in effect during the period beginning with the installation of the friable, fiber-releasing asbestos materials and ending when those materials are removed or contained. In deciding to apply an equitable continuous trigger, the appellate court was persuaded by the analysis in *United States Gypsum Co. v. Admiral Insurance Co.*[10] i.e., that property damage, whether from the presence of airborne asbestos fibers or settled fibers subject to reentrainment, occurs over a span of time and cannot be linked to or confined to different policy periods and that the difficulty in determining precisely when the damage occurred or in following its progress or proliferation is exacerbated by the interchangeable nature of the fibers and that their release occurs on a continuing basis.

Determining when property damage from airborne mold spores occurs is equally difficult for the same reasons: the interchangeable nature of the spores and that their release occurs on a continuing basis. *The Board of Education of Township High School District No. 211* case supports not only the position that mold spore contamination constitutes physical loss or damage,[11] but also the adoption of the equitable continuous trigger to determine the timing of mold damage for coverage purposes, beginning with the initial water event that caused the ensuing mold damage and ending when the moldy building materials are removed

The Illinois Supreme Court in American States Insurance Co. v. Koloms[12] concluded that the "pollution" exclusion applies only to injury or damage caused by traditional environmental pollution i.e., industrial discharge of toxic waste into the environment.[13] Accordingly, even if it is considered a "pollutant," neither mold growth nor the discharge, dispersal, migration, release, or escape of mold spores in a contained or indoor environment, constitutes the type of traditional environmental pollution contemplated by the exclusion.

Illinois courts adhere to the rule of insurance contract construction known as *noscitur a sociis*, which requires that language be construed in accordance with the words which are associated with it.[14] The term "mold" in the typical "mold" exclusion is surrounded by types of loss (rust, corrosion, wet or dry rot) that occur naturally over a period of time. As such, the intent of the mold exclusion is to bar coverage for mold which occurs gradually, and which is not associated with a fortuitous event. This also is the interpretation advanced by the Illinois Department of Insurance on its website as well as in response to inquiries made by insureds.[15] Interpreting the mold exclusion in the manner above is not only consistent with but required by the well-established rule of insurance contract construction in Illinois that "[p]rovisions in insurance policies that limit or exclude coverage are to be construed liberally in favor of the insured."[16] Under this canon of insurance contract construction, even if a particular limiting or exclusionary phrase or term is capable of being interpreted in the manner sought by the insurer, where another interpretation favorable to the insured reasonably can be made, that construction must be applied. Interpreting the mold exclusion liberally in favor of insureds thus results in the conclusion that mold damage which originates from a covered peril is not excluded.

In addition to the "mold" and "pollution" exclusions, many, if not most, first-party property insurance policies exclude loss or damage caused by faulty, inadequate, or defective: (1) workmanship, repair, construction, renovation, or remodeling; (2) materials used in repair, construction, renovation, or remodeling; or (3) maintenance. However, the "faulty, inadequate or defective" exclusion typically covers any "ensuing loss" not excluded or excepted by the policy. In *Board of Education of Maine Township High School District 207 v. International Insurance Co.*, the appellate court interpreted the ensuing loss exception as follows: "if a nonexcluded loss is caused by perils that either occur after an excluded peril or are causally related to an excluded peril, the nonexcluded loss will remain covered. Only the nonexcluded portion of the loss will be covered, however. The uncovered peril itself will not be covered."[17]

As an example, a roof leaks, with resultant interior water and mold damage. The roof was faultily maintained. Under the interpretation espoused by the appellate court in *Board of Education of Maine Township*, the leaking roof itself is not covered because of the "faulty maintenance" exclusion; but, the interior water and mold damage is covered as an "ensuing loss" because (a) the standard "water damage" exclusion does not preclude coverage for interior water damage caused by rain or melting ice or snow and (b) the mold damage is the result or consequence of a covered water peril.[18]

In Illinois, in addition to breach of contract, causes of action available to an insured against his or her property insurer in a mold damage case include: (1) negligence; (2) consumer fraud; (3) violation of section 155 of the Illinois Insurance Code; and (4) equitable relief, including injunctive and declaratory relief, as well as specific performance.

Negligence actions may be available to an insured in a mold damage case if the remediation goes sour. Because of the lack of regulatory or professional standards as to acceptable or unacceptable mold contamination levels, a "successful" remediation may be subject to dispute. However, common sense dictates that a "successful" remediation is one in which the insured is able to re-occupy his or her premises without health complaints or physical discomfort.

A negligence action may be available against the insurer after an unsuccessful remediation if the remediation contractor was acting as the insurer's agent. A principal-agent relationship can exist through a contract or guaranteed contractor program. It also can exist if the insurer (1) directed, supervised, and managed the remediation contractor, and exercised decision-making authority as to the scope of the mold remediation and (2) paid the remediation contractor directly.

Typical negligent acts or omissions on the part of a remediation contractor in cleaning, restoring, and/or remediating the insured premises and personal property include:

- failing to dry the insured premises;
- failing to take proper measures to protect the insured premises from continued water intrusion;
- failing to take reasonable steps to prevent or to inhibit the microbial growth in the insured premises;
- failing to warn the insured of the dangers and biohazards it left behind in the insured premises;
- failing to properly address the water damage and the dangerous microbial contamination that ensued;
- failing to complete the cleaning, restoration, and remediation of the insured premises;
- failing to restore the insured premises to its pre-loss condition;
- failing to perform the cleaning, restoration, and remediation services in a good and workmanlike manner;
- failing to perform the cleaning, restoration, and remediation services according to established guidelines or standards for fungal or mold problems; and
- neglecting, mistreating, and otherwise damaging the personal property and household goods and contents by, among other things, failing to take reasonable steps to clean it, failing to adequately store it, and otherwise failing to take steps to protect it from additional loss and damage.

A negligence action also may be available against the insurer after an unsuccessful remediation if the insurer voluntarily undertook to direct, supervise, manage, and employ experts or specialists to inspect, assess, remediate, and contain the mold growth in the insured premises. Typical negligent acts or omissions on the part of the insurer include:

- allowing claim personnel who have no experience or training in mold to direct, supervise, manage, and employ an expert or specialist, such as a certified industrial hygienist, to inspect the insured premises to determine if there is mold growth in the insured premises;

- allowing claim personnel who have no experience or training in mold to direct, supervise, manage, and employ a remediation contractor to inspect, assess, remediate, and contain the mold growth in the insured premises;
- allowing claim personnel who have no experience or training in mold to exercise decision-making authority as to the scope of diagnostic testing, sampling, and investigation to be performed by the expert or specialist;
- allowing claim personnel who have no experience or training in mold to exercise decision-making authority as to what cleaning, restoration, remediation, repair, replacement and/or renovation or other services will or will not be performed by the remediation contractor;
- hiring a remediation contractor who is not qualified to inspect, assess, remediate, and contain the mold growth in the insured premises;
- failing to direct the remediation contractor to complete the cleaning, restoration, and remediation of the insured premises;
- failing to direct the remediation contractor to clean, restore, and remediate the insured's personal property and household goods and contents;
- failing to properly supervise the remediation contractor so as to ensure that its cleaning, restoration, and remediation of the insured premises was performed in a good and workmanlike manner;
- failing to properly supervise the remediation contractor so as to ensure that its cleaning, restoration, and remediation of the insured premises was performed according to established industry guidelines or standards;
- failing to direct that a final "clearance" test be conducted of the insured premises so as to ensure (i) that the remediation work has eliminated the mold growth in the insured premises and (ii) that the insured premises is fit for human occupation;
- failing to direct the expert or specialist to collect samples of building materials and to analyze for mold content; and

- failing to direct the expert or specialist to collect samples of dust/debris from the personal property and household goods and contents in the insured premises and to analyze for mold content so as to determine if any remediation of the personal property and household goods and contents is necessary.

Amica Mutual Insurance Co. v. Henderson is a case which illustrates an insurer's, a remediation contractor's, and an environmental testing company's exposure arising out of an unsuccessful remediation. In that case, the insured's allegations were as follows. Water and mold damage to her home was discovered following a fire loss. Her homeowners insurer retained service companies to repair and to remediate the damage to her property. The repair and remediation work was performed negligently. As a result, water was able to seep into her home, causing toxic mold to grow and to spread throughout the home. Her homeowners insurer retained specialized service companies to perform investigations and microbial toxin testing at her property. The district court denied the environmental testing company's motion to dismiss. The district court concluded that the insured had sufficiently pled negligent conduct on the part of the environmental testing company (failing to tell the insured that the mold in the home could be toxic, that contact with the mold could be hazardous, that the remediation contractor's work did not isolate and contain the mold contamination, and allowing her and her family to reside in the home knowing the toxicity of the mold) and that this negligent conduct caused the insured and her children emotional distress.[19]

The Illinois Consumer Fraud and Deceptive Business Practices Act (Consumer Fraud Act) states, in pertinent part, that "[u]nfair methods of competition and unfair or deceptive acts or practices, including but not limited to the use or employment of any deception, fraud, false pretense, false promise, misrepresentation or the concealment, suppression or omission of any material fact … are hereby declared unlawful."[20] The act was intended to protect consumers against fraud, unfair methods of competition, and unfair or deceptive acts or practices in the conduct of trade or commerce.[21] The Consumer Fraud Act is to be broadly construed to prohibit deception or false promises. As Illinois courts have stated, "we perceive a clear mandate from the Illinois legislature that the courts of this

State are to utilize the Consumer Fraud Act to the utmost degree in eradicating all forms of deceptive and unfair business practices and to grant appropriate remedies to defrauded consumers."[22] Because the Consumer Fraud Act was designed to provide broader protection for consumers than common law fraud, a plaintiff is not required to prove every element of common law fraud in order to recover. In order to state a claim for a violation of the Consumer Fraud Act, a plaintiff must allege the following: (1) a deceptive act or practice; (2) an intent by defendant that plaintiff rely on the deception; and (3) that the deception occurred in the course of conduct involving trade or commerce.[23]

Deceptive marketing practices are actionable as independent torts, provided the deceptive marking or advertising actually deceives the insured.[24] In that regard, mold damage claims can be appropriate Consumer Fraud Act claims. For example, one insurer's marketing slogans are to get its insureds back to where they belong and to restore their lives to order following a loss.[25] Its failure to do so, particularly after an unsuccessful remediation where it directs, supervises, and manages the remediation contractor and exercises decision-making authority as to the scope of the remediation, could constitute a deceptive act or practice within the meaning of the Consumer Fraud Act.

Section 155 of the Illinois Insurance Code "provides an extracontractual remedy to policyholders whose insurer's refusal to recognize liability and pay a claim under a policy is vexatious and unreasonable."[26] Section 155 allows as part of the taxable costs in an action reasonable attorney fees and an amount not to exceed $60,000.[27]

Examples of vexatious and unreasonable conduct on the part of the insurer include: (a) refusing to pay despite knowledge that it was liable and knowledge that its refusal would result in serious damage to the insured; (b) failing to adequately investigate a claim; (c) denial of the claim without adequate supporting evidence; (d) failing to evaluate a claim objectively; and (e) interpreting policy provisions in an unreasonable manner.[28]

The question of vexatious and unreasonable conduct within the meaning of section 155 is a factual one, which must be based upon an assessment of the totality of the circumstances, including: the insurer's attitude; whether an insured was forced to file suit to recover; and if an insured was deprived of the use of his or her property.[29]

Vexatious and unreasonable conduct on the part of an insurer in a mold damage claim which, if proven, should constitute a violation of section 155 includes:

- failing to promptly contact the insured once the loss is reported;
- failing to pay the insured amounts due under the insurance policy within forty days of the loss, which constitutes an unreasonable delay in paying the insured's claim as a matter of law;
- failing to provide the insured with a reasonable written explanation for the delay in resolving his or her claim, in violation of the regulations promulgated by the Illinois Director of Insurance;
- adjusting the claim contrary to and in violation of its own water and mold adjusting guidelines;
- adjusting the claim contrary to and in violation of its own property claims handling procedures;
- assigning claim representatives to adjust the claim who are untrained and inexperienced in mold, and who make numerous misrepresentations to the insured;
- not attempting in good faith to effectuate a prompt, fair, and equitable settlement of the claim, a claim in which liability is reasonably clear;
- without proper cause, wrongfully and knowingly refusing to reimburse the insured for certain losses, and delaying payment on other losses which it acknowledges are covered under the insurance policy, such losses including additional living expenses and other structural repairs and replacements;
- forcing the insured to retain legal counsel to investigate his or her claim and to file a lawsuit to recover the benefits that should have been immediately forthcoming under the insurance policy;
- failing to acknowledge with reasonable promptness pertinent communications with respect to the claim;

- failing to provide the insured with reports prepared by experts, despite repeated requests for the same;
- willfully refusing to undertake its coverage obligations to protect the insured, instead undertaking a continuous effort to conceal material information from the insured, to procrastinate and to delay at every turn, and acting to maximize its own financial interest at the expense and to the financial demise of the insured;
- denying the insured's claim without adequate supporting evidence;
- denying the insured's claim based on an unreasonable and erroneous interpretation of its own insurance policy provisions; and
- denying the insured's claim knowing that it was liable and that there was no bona fide or genuine dispute as to coverage for the water damage and mold damage.

The construction of an insurance contract is a proper subject of inquiry in a declaratory judgment action.[30] Declaratory relief is thus appropriate in a mold damage case if there is a dispute over coverage. Specific performance may be appropriate in a mold damage case if the insurer fails to honor specific agreements made with its insured, such as agreeing to pay for the following items: the cost to pack, to move, and to store the personal property and household goods and contents inside the insured premises while the insured premises is being remediated and reconstructed; the cost to clean the insured's clothing; the cost of the mold remediation work to be performed at the insured premises; the cost to reconstruct the interior of the insured premises following the completion of the mold remediation work; and all of the increased living expenses necessarily incurred by the insured while the insured premises is uninhabitable and unfit for human occupation.[31] Injunctive relief is appropriate in a mold damage case if: the insurer is threatening to cut-off the insureds' additional living expenses prior to the completion of remediation and reconstruction (such as rent and rental furniture, if the insured is required to move out of the insured dwelling) and the insureds are financially unable to pay these expenses themselves.[32]

6.3 Residential Real Property Disclosure Act

The Illinois Residential Real Property Disclosure Act requires sellers of residential real estate to provide potential buyers with a written disclosure of defects and other conditions of the property. Included in the report are such items as flooding or recurring leakage, asbestos, termites, underground tanks and, in some realtor disclosure reports, mold.[33] The act provides that a seller is not liable for any inaccuracy, omission or error in the report if: (1) he or she had no knowledge of such error, inaccuracy or omission; (2) he or she had a reasonable belief that such error, inaccuracy, or omission was correct; or (3) the error, inaccuracy, or omission is based on information provided by a public agency or licensed professional, and the seller had no knowledge of such error, inaccuracy, or omission.[34] Sellers that knowingly violate this act and fail to truthfully complete the disclosure report shall be liable for actual damages, court costs, and reasonable attorney fees.[35]

Claims against a seller of residential real estate for violation of the act, fraudulent misrepresentation, and negligent misrepresentation in failing to disclose known material defects in the property resulting in water infiltration and ensuing mold damage, may not be covered under the seller's homeowners liability insurance coverage, as was the case in *Allstate Insurance Co. v. Lane*.[36] In that case, the buyers sought rescission of a real estate contract and damages because the home they purchased from the sellers had been damaged by extensive water infiltration before the closing. The buyers alleged that the sellers replaced most of the windows in the house; but, because of improper installation, the windows leaked and rainwater entered the house. The buyers claimed that the sellers knew of the water infiltration when they discovered damaged plaster, bubbling in the wallpaper, and puddling in the basement during heavy rains, but misrepresented the same when they signed a residential real property disclosure report stating that they were not aware of either recurring leaks in the basement, material defects in walls or floors, or leaks or material defects in the roof, ceiling or chimney. The sellers' Allstate Family Liability insurance policy provided that "Allstate will pay damages which an insured person becomes legally obligated to pay because of bodily injury or property damage arising from an occurrence to which this policy applies, and is covered by this part of the policy." The appellate court concluded that Allstate was not obligated either to defend or to indemnify the sellers in the underlying lawsuit because the buyers' claims did not arise from an "occurrence" within the meaning of the Allstate policy. The appellate court reasoned that the actual accident or occurrence was the faulty installation of windows and water infiltration, leading to the structural deterioration of the house. The buyers' claim pertained to the nondisclosure of the damage, not to the damage itself. Accordingly, the underlying claims were outside the scope of the Allstate policy.

6.4 Personal Injury and Property Damage Claims

In Illinois, the success of a mold-related property damage tort claim requires pleading around the economic loss doctrine. First enunciated in Illinois in 1982 by the Supreme Court in *Moorman Manufacturing Co. v. National Tank Co.*,[37] the doctrine precludes the recovery in tort of purely economic losses i.e., economic losses unaccompanied by property damage or personal injury resulting from a sudden, calamitous, or dangerous event.

Toxic contamination of a building itself, other parts of a building, and the building contents is damage to "other property," as seen in *Board of Education of City of Chicago v. A, C, and S, Inc.*[38] There, the issue was whether the plaintiffs, thirty-four school districts, had sufficiently pleaded negligence and strict liability causes of action to recover the removal and repair costs of asbestos-containing material (ACM) in their buildings from various defendants involved at some level of the manufacturing and distribution chain of ACM. The linchpin for both of these causes of action was whether the complaints sufficiently alleged that asbestos had caused damage to other property or injury to persons so as to fall within a tort claim, as opposed to a contract cause of action. The defendants contended that the plaintiffs only suffered economic loss, which is recoverable in contract and not in tort. They argued that there was alleged only a risk of harm to people and no actual harm was pleaded. Further, they argued that no property damage was alleged in the complaint.

The Illinois Supreme Court disagreed, holding that the complaints alleged sufficient facts to establish tort actions under the principles established in *Moorman*. The complaints alleged that friable asbestos existed in plaintiffs' buildings and that asbestos had been released throughout the schools. The Court stated that "it would be incongruous to argue there is no damage to other property when a harmful element exists throughout a building or an area of a building which by law must be corrected and at trial may be proven to exist at unacceptably dangerous levels."[39] Relying on cases in which courts had found that asbestos fibers may contaminate a building sufficiently to allege damage to property, the Court concluded that the complaints alleged damage to property because the buildings were contaminated with a toxic substance which rendered them unsafe for their normal use.

A, C & S, Inc. is not limited to asbestos contamination. In *American United Logistics, Inc. v. Catellus Development Corp.*,[40] the Seventh Circuit Court of Appeals concluded that recovery for contamination is not barred by the economic loss doctrine if the contamination is hazardous or unreasonably dangerous. Accordingly, this author submits that, under *A, C & S, Inc. and American United Logistics, Inc.*, if the toxic microbial contamination inside a dwelling or building is hazardous or unreasonably dangerous i.e., the toxic microbial contamination poses a serious health risk to the occupants, then all losses caused by the contamination, including economic losses, are recoverable.

In *Murfield Village-Vernon Hills, LLC v. Reinke*,[41] it was alleged that wet or saturated insulation or carpentry materials were negligently installed during construction of a home, which led to the growth of unacceptably high levels of mold and bacteria. While the growth of the mold and bacteria occurred gradually, the appellate court, nonetheless, concluded that its manifestation was "sudden and calamitous," damaging the homeowners' personal property and requiring them to flee their home or experience the likelihood of personal injury. Because the occurrence was sufficiently sudden and calamitous to place it under the exception to the economic loss rule for property damage resulting from a sudden or dangerous occurrence, the appellate court allowed recovery in tort for the loss to the homeowners' personal property. However, as to the damages accruing from the loss of the home, the appellate court barred recovery in tort, reasoning that it was a "classic" economic loss. This author submits that the *Reinke* decision is inconsistent with *Moorman* and its progency, which establish that a tort claim which alleges economic loss in combination with non-economic loss (personal injury and/or property damage) is not barred by the economic loss doctrine.[42]

The premise behind the economic loss doctrine is that tort law is appropriately suited for personal injury or property damage resulting from a sudden or dangerous occurrence; whereas, contract and warranty law are appropriately suited for loss relating to a purchaser's disappointed expectations when a defect renders a product (which can be a home under Illinois law) inferior or unable adequately to perform its intended function. With respect to warranty law, the Illinois Supreme Court held in *Peterson v. Hubschman Construction Co.*[43] that, in the sale of a new house by a builder-vendor, there is an implied warranty of habitability which will support an action against the builder-vendor by the original purchaser.[44] To establish a cause of action for breach of the implied warranty of habitability, the plaintiff must plead and prove that the home had a latent defect caused by improper design, material, or workmanship in its construction which rendered the property unsuitable for use as a home. A latent defect "is one which, in the circumstances of the case, would not have been discovered by the exercise of ordinary and reasonable care."[45] The remedy for breach of the implied warranty of habitability is rescission if the defects were discovered before the conveyance to the purchasers had been completed, or damages measured by the cost of correcting the defective conditions.[46]

Since *Peterson*, the implied warranty of habitability has been extended by Illinois courts to include: actions against a developer-vendor of a new condominium,[47] actions against a builder and/or developer-vendor by subsequent purchasers, provided the latent defects manifest themselves within a reasonable time after the purchase of the home;[48] actions against subcontractors of a builder-vendor where the builder-vendor has been dissolved as an entity and is insolvent;[49] and actions by subsequent purchasers against builders that construct significant structural additions to an existing residence.[50]

The success of a mold-related personal injury claim usually comes down to the admissibility of expert testimony with respect to causation of illness and injury by mold. In Illinois, the admissibility is governed by what is known as the Frye[51] standard, which standard dictates that scientific evidence is admissible at trial only if the methodology or scientific principle upon which the opinion is based is sufficiently established to have gained general acceptance in the particular field in which it belongs. In this context, "general acceptance" does not mean universal acceptance, and it does not require that the methodology in question be accepted by unanimity, consensus, or even a majority of experts. Instead, it is sufficient that the underlying method used to generate an expert's opinion is reasonably relied upon by experts in the relevant field. Significantly, the Frye test applies only to "new" or "novel" scientific methodologies.[52] There is no reported appellate decision in Illinois regarding the admissibility of expert medical testimony in mold-related personal injury cases.

In Illinois, any mold-related personal injury or property damage action, whether based upon tort, contract, or otherwise,[53] arising out of an act or omission in the design, planning, supervision, observation, or management of construction, or construction of an improvement to real property, must be commenced within four years from the time the person bringing the action knew or should reasonably have known of such act or omission.[54] Additionally, any mold-related personal injury or property damage action not brought within ten years of an act or omission in the design, planning, supervision, observation, or management of construction, or construction of an improvement to real property is barred; unless, the act or omission is discovered prior to the expiration of the ten-year period of repose, in which case the action must be brought within four years from the date of discovery.[55]

The limitation and repose periods articulated above do not apply to bar any action against a defendant who has expressly warranted or promised the improvement to real property for a longer period.[56] However, many new construction contracts expressly disclaim the implied warranty of habitability and/or limit the period of its express warranty to one year from the date of closing. A disclaimer is a bar to the cause of action for breach of the implied warranty of habitability if: it references the implied warranty of habitability by name; it is a conspicuous provision which fully discloses the consequences of its inclusion; and the agreement of the parties in fact included the disclaimer.[57]

6.5 Third-Party Liability Insurance

A third-party liability insurance policy provides coverage for the insured's legal liability to another arising from claims resulting in "bodily injury" or "property damage" which occur during the policy period. Liability coverage issues involving mold-related bodily injury or property damage caused by water or moisture intrusion resulting from faulty workmanship or defective construction include whether the "bodily injury" or "property damage" was caused by an "occurrence" and whether the injury or damage occurred during the policy period.

"Occurrence" is typically defined in liability insurance policies as "an accident, including continuous or repeated exposure to substantially the same general harmful conditions." The term "accident" is usually an undefined term. The dictionary definition of "accident" is "an event or condition occurring by chance or arising from unknown or remote causes."[58] *Black's Law Dictionary* defines "accident" as follows: "The word 'accident,' in accident policies, means an event which takes place without one's foresight or expectation. A result, though unexpected, is not an accident; the means or cause must be accidental."[59] Illinois courts have concluded that damage caused by faulty workmanship or defective construction standing alone is not an "occurrence."[60]

Not only must there by an "occurrence," general liability insurance policies require that bodily injury and/or property damage occur during the policy period. Simply stated, the insurance policy in effect at the time injury or damage occurs is the policy that "triggers" coverage, even if a claim is not made until after the policy expires. *The Board of Education of Township High School District No. 211 v. International Insurance Co.* case previously discussed supports adoption of the equitable continuous trigger to determine the timing of mold damage resulting from water intrusion caused by faulty workmanship or defective construction, beginning with the initial water event that caused the ensuing mold damage and ending when the moldy building materials are removed.

6.6 Mold Legislation

In June, 2003, the 93rd General Assembly of the State of Illinois passed House Joint Resolution 12. This Resolution created the Joint Task Force on Mold in Indoor Environments. In creating this Joint Task Force, the General Assembly recognized that:

> "The presence of mold in indoor environments presents a public health risk to the citizens of Illinois;"
>
> "Mold in schools, places of employment, and homes has been shown to adversely affect the health of Illinois' children and senior citizens," and

> "The adverse health effects of exposure to molds on the general population and the specific effects of mold on certain subpopulations, including infants, children, the elderly, pregnant women, asthmatics, allergic individuals, immune-comprised individuals, and others are resulting in illness and increased medical costs."

The purpose of the Joint Task Force is "examining the mold issue in Illinois and making recommendations to the General Assembly pertaining to regulation of mold in indoor environments in Illinois."

In February, 2004, the 93rd General Assembly of the State of Illinois introduced the Toxic and Pathogenic Mold Protection Act. The act, which has yet to become law, requires the Illinois Department of Public Health: to consider adopting permissible exposure limits for molds; to adopt standards for reducing the occurrence of mold contamination in indoor environments; to adopt mold identification guidelines; and to develop mold abatement standards. The act also requires a seller of residential, commercial, or industrial real property to disclose to prospective buyers the existence of mold in the indoor environment. Finally, the act authorizes enforcement of the Department's standards and guidelines by local health authorities and others.

In December, 2004, the Joint Task Force issued its final report of "Mold in Indoor Environments." The report summarized the activities of the Task Force and made various recommendations, including (1) that parties providing mold remediation services in residential, public, and commercial buildings in Illinois be required to register with the Illinois Department of Public Health and provide proof of financial responsibility and (2) that laboratories performing tests to confirm mold contamination be certified by the American Industrial Hygiene Association using nationally recognized accreditation standards set by the Environmental Microbiology Laboratory Accreditation Program.

Endnotes

1. *Michael and Juanita Jordon v. STO Corporation, et al.*, Circuit Court of Cook County, Illinois No. 04 L 3097. The essence of the lawsuit is that the STO exterior insulation and finish system ("EIFS") cladding applied to the Jordans' home failed to provide for the collection and redirection of foreseeable water that intrudes behind the cladding; instead, the EIFS traps water between the

cladding, the structural components, and interior finishes of building walls and, as a consequence, results in deterioration to structural components, ensuing damage to interior finishes and other building components, and the growth and proliferation of toxic microbial contamination throughout the home. The complaint seeks to recover the cost to remove and to replace the EIFS cladding under contract and warranty theories.

2. According to the National Association of Mutual Insurance Companies, "[w]hen it comes to the insurance industry, mold is not part of the standard insurance policy and is covered only if it's a direct result of a covered peril" (www.moldupdate.com). MoldUpdate.com, which is owned and operated by the National Association of Mutual Insurance Companies, is a site where those interested in the issue of mold can come for current news, education, scientific links, litigation updates, state legislation, and matters of mold as it relates to the insurance industry.

3. *See* Illinois Department of Insurance Bulletin CB#2002-07, dated December 5, 2002.

4. 215 ILL. COMP. STAT. 5/397 (2004); 50 Ill Adm. Code § 2301.100 (effective March 17, 1961).

5. *Lundquist v. Allstate Insurance Co.*, 314 Ill.App.3d 240 (2000).

6. Standard Fire Policy, Insuring Clause.

7. Standard Fire Policy, Lines 14-24.

8. In *Wallis v. Country Mutual Insurance Co.*, 309 Ill.App.3d 566, 570 (2000), the appellate court articulated the respective burdens of proof in a lawsuit for an insurer's failure to pay a claim under an all-risk insurance policy. The insured's burden is to establish (1) that a loss occurred, (2) that the loss resulted from a fortuitous event, and (3) that an all-risk policy covering the insured property was in effect at the time of loss. Once an insured establishes a *prima facie* case, the burden then shifts to the insurer to show that the loss resulted from a peril expressly excluded from coverage.

9. 308 Ill.App.3d 597 (1999).

10. 268 Ill.App.3d 598 (1994).

11. A variety of contaminating conditions have been held to constitute "physical loss of or damage to property." See, e.g., *Matzner v. Seaco Insurance Co.*, 1998 WL 566658 (Mass.Super. Aug. 12, 1998) (carbon monoxide contamination);

Columbiaknit, Inc. v. Affiliated FM Insurance Co., 1999 WL 619100 (D.Or. Aug. 4, 1999) (mildew contamination); *Farmers Insurance Co. of Oregon v. Trutunich*, 858 P.2d 1332 (1993) (odors from illegal methamphetamine cooking).

12. 177 Ill.2d 473 (1997).

13. See also *Kim v. State Farm Fire and Casualty Co.*, 312 Ill.App.3d 770 (2000).

14. *Hardware Wholesaler, Inc. v. Heath*, 10 Ill.App.3d 337, 343 (1973). *See also Black's Law Dictionary 1660* (6th ed. 1990).

15. "Illinois Insurance Facts Mold Facts and Homeowners Insurance," dated December, 2002 at http://www.ins.state.il.us/HomeInsurance/Mold.htm.

16. *Allstate Insurance Co. v. Davenport*, 309 Ill.App.3d 750, 754 (1999) (construing the term "continuously" under an "any occupation" disability provision of an automobile insurance policy liberally in favor of the insured).

17. 292 Ill.App.3d 14, 21 (1997).

18. This is the interpretation advanced by the Property Loss Research Bureau ("PLRB"), which is a recognized resource in the insurance industry for insurers in interpreting policy provisions. See PLRB Question & Answer Reference Service 146.1, "Interior Water Damage" (July, 1996). *See also Phillips v. United Services Automobile Association*, 146 S.W. 3d 629 (Tenn. Ct. App. 2004); *Arnold v. Cincinnati Insurance Co.*, 276 Wis.2d 762 (Wis. App. 2004).

19. No. 02 C 5193, 2003 WL 21196261 (N.D. Ill. May 15, 2003).

20. 815 ILL. COMP. STAT. 505/2 (2004).

21. *Washington Courte Condominium Assocation-Four v. Washington-Golf Corp.*, 267 Ill.App.3d 790 (1994).

22. *American Buyers Club of Mt. Vernon, Illinois, Inc. v. Honecker*, 46 Ill.App.3d 252, 257 (1977).

23. *Connick v. Suzuki Motor Co. Ltd.*, 174 Ill.2d. 482, 501 (1977); *Siegel v. Levy Organization Development Co., Inc.*, 153 Ill.2d 534 (1997).

24. *See Shannon v. Boise Cascade Corp.*, 208 Ill.2d 517 (2004). In that case, homeowners failed to establish that a composite siding manufacturer's deceptive advertising was a proximate cause of their damages under the Consumer

Fraud Act because there was no allegation that the homeowners were actually deceived by the manufacturer's allegedly deceptive advertising, or that product literature had deceived a particular builder, architect, or contractor, resulting in installation of defective siding on the homes. *See also Commonwealth Insurance Co. v. Stone Container Corp.*, No. 99 C 8471, 2001 WL 477151 (N.D. Ill. May 3, 2001) and *Industrial Enclosure Corp. v. Northern Insurance Corp. of New York*, No. 97 C 6850, 1998 WL 852845 (N.D. Ill. 1998).

25. This is the Farmers Insurance Group marketing slogan, which is set forth in its website materials at http://www.farmersinsurance.com/.

26. *Cramer v. Insurance Exchange Agency,* 174 Ill.2d 513, 516 (1996); *see also Richardson v. Illinois Power Co.,* 217 Ill.App.3d 708, 715 (1991) ("The legislature intended to provide a remedy to an insured who encounters unnecessary difficulties when an insurer withholds policy benefits."); *Myrda v. Coronet Insurance Co.*, 221 Ill.App.3d 482, 491 (1991) ("[S]ection 155 of the Code is intended to aid the insured and to encourage insurers from profiting by their superior financial positions while delaying in the payment of contractual obligations."); *McGee v. State Farm Fire and Casualty Co.*, 315 Ill.App.3d 673, 681 (2000) ("The attorney fees, costs, and limited penalty provisions of section 155 are intended to make suits by policyholders economically feasible and punish insurance companies for misconduct.").

27. 215 ILL.COMP.STAT. 5/155 (2004).

28. *Emerson v. American Bankers Insurance Co.*, 223 Ill.App.3d 929, 936 (1992).

29. *Mohr v. Dix Mutual Country Fire Insurance Co.*, 143 Ill.App.3d 989, 998-99 (1986).

30. *Farmers Automobile Insurance Association v. Janusick*, 30 Ill.App.2d 352 (1961).

31. *See, e.g., New Park Forest Associates II v. Rogers Enterprises, Inc.*, 195 Ill.App.3d 757 (1990).

32. *Lamka v. Illinois Farmers Insurance Co.*, Circuit Court of Cook County, Illinois No. 02 CH 4388; *Cianci v. Safeco Insurance Co. of Illinois*, Circuit Court of Cook County, Illinois No. 02 CH 4670.

33. 765 ILL. COMP. STAT. 77/20 (2004).

34. 765 ILL. COMP. STAT. 77/35 (2004).

35. 765 ILL. COMP. STAT. 77/25 (2004).

36. 345 Ill.App.3d 547 (2003).

37. 91 Ill.2d 69 (1982).

38. 131 Ill.2d 428 (1989).

39. *Id.* at 446.

40. 319 F.3d 921 (7th Cir. 2003).

41. 394 Ill.App.3d 178 (2004).

42. See *In re Chicago Flood Litigation*, 176 Ill.2d 179, 201 (1997) ("Absent injury to a plaintiff's person or property, a claim presents an economic loss not recoverable in tort."); *Mars, Inc. v. Heritage Builders of Effingham, Inc.*, 327 Ill.App.3d 346, 351 (2002) ("A plaintiff may still recover economic damages in tort when the plaintiff has sustained a personal injury or property damage as a result of a sudden or dangerous occurrence."); *Great Oak, L.L.C. v. Begley Co.*, 2003 WL 880994 (N.D. Ill. March 5, 2003) ("[T]he pleadings sufficiently allege damage to property to overcome the Moorman doctrine. Although it appears that the bulk of Plaintiff's damages are in the nature of economic losses, the complaint alleges damages to Plaintiff's property as well as clean up damages and future damages.") *See also Daanen & Janssen, Inc. v. Cedarapids, Inc.*, 216 Wis.2d 395, 573 N.W.2d 842, 845 (1998) ("The economic loss doctrine, however, does not bar a commercial purchaser's claim based on personal injury or damage to property other than the product, or economic loss claims that are alleged in combination with noneconomic losses.").

43. 76 Ill.2d 31 (1979).

44. The implied warranty of habitability also applies to all leases of residential real estate. *Glasoe v. Trinkle*, 107 Ill.2d 1 (1985). In order to assert a claim for a landlord's breach of the implied warranty of habitability, a tenant must show either that he or she gave notice of the alleged defect to the landlord or that the landlord had knowledge of the alleged defect. *Abram v. Lithman*, 150 Ill.App.3d 174 (1986). The remedy for a landlord's breach of the implied warranty is the deterioration in the benefits from the lease rather than for personal injuries or property damage. *Auburn v. Amoco Oil Co.*, 106 Ill.App.3d 60 (1982).

45. *Ollivier v. Alden*, 262 Ill.App.3d 190, 197 (1994).

46. *Park v. Sohn*, 89 Ill.2d 453 (1982).

47. *Tassan v. United Development Co.*, 88 Ill.App.3d 581 (1980).

48. *Redarowicz v. Ohlendorf*, 92 Ill.2d 171 (1982).

49. *Minton v. Richards Group of Chicago*, 116 Ill.App.3d 852 (1983). Where a Plaintiff timely files an action against a general contractor for faults or defects in construction, and the general contractor subsequently becomes insolvent allowing the Plaintiff to bring an action against a subcontractor for breach of implied warranty of habitability under *Minton*, the statute of limitations is triggered at the time that the Plaintiff knew or reasonably should have known of the general contractor's insolvency. *Dearlove Cove Condominiums v. Kin Construction Co., Inc.*, 180 Ill.App.3d 437 (1989).

50. *VonHoldt v. Barba & Barba Construction, Inc.*, 175 Ill.2d 426 (1997). In this case, the Illinois Supreme Court also held that the reasonable time for bringing a cause of action under the "reasonable time" standard of *Redarowicz* begins with the act or omission causing the defect rather than the date on which the subsequent purchaser takes title to the property.

51. *Frye v. United States*, 293 F.1013 (D.C. Cir. 1923).

52. *Donaldson v. Central Illinois Public Service Co.*, 199 Ill.2d 63 (2002). In *In re commitment of Stephen E. Simons*, 213 Ill.2d 523 (2004), the Illinois Supreme Court adopted a dual standard of review with respect to the trial court's admission of expert scientific testimony. The decision as to whether an expert scientific witness is qualified to testify in a subject area, and whether the proffered testimony is relevant in a particular case, remains in the sound discretion of the trial court. The trial court's *Frye* analysis, however, is subject to *de novo* review. In conducting such *de novo* review, the reviewing court may consider not only the trial court record but also, where appropriate, sources outside the record, including legal and scientific articles, as well as court opinions from other jurisdictions.

53. Several Illinois courts have applied section 13-214 of the Illinois Code of Civil Procedure to actions against persons involved in construction-related activities based on allegations of the breach of the implied warranty of habitability. *See, e.g., Eickmeyer v. Blietz Organization, Inc.*, 284 Ill.App.3d 134 (1996); *Schleyhahn v. Cole*, 178 Ill.App.3d 111 (1989); *Zielinski v. Miller*, 277 Ill.App.3d 735 (1995).

54. 735 ILL. COMP. STAT. 5/13-214 (a) (2004).

55. 735 ILL. COMP. STAT. 5/13-214 (b) (2004). The time of accrual under this section governing a purchaser's suit against a builder to recover for latent defects in the purchaser's new house begins from the date the purchaser takes possession of the house. *See Andreoli v. John Henry Homes, Inc.*, 297 Ill.App.3d 151 (1998).

56. 735 ILL. COMP. STAT. 5/13-214 (d) (2004).

57. In *Board of Managers of Chestnut Hills Condominium Association v. Pasquinelli*, 354 Ill.App.3d 749 (2004), the disclaimer was found to be invalid because the record in the case did not show an effective waiver of the implied warranty of habitability. There was no evidence that the builder-vendor called the purchasers' attention to the disclaimer provision, or instructed the purchasers to specifically initial the provision. The provision also failed to set forth the consequences of the waiver of the implied warranty of habit ability. In *Board of Managers of the Village Centre Condominium Association, Inc. v. Wilmette Partners*, 198 Ill.2d 132 (2001), the disclaimer also was found to be invalid because it did not refer to the implied warranty of habitability by name. In *Breckenridge v. Cambridge Homes, Inc.*, 246 Ill.App.3d 810 (1993), the appellate court found the disclaimer to be valid because it was in a conspicuous location, in large-size and bold print, and in plain language. The Plaintiffs also initialed the disclaimer provision indicating that they had accepted its terms.

58. *Webster's Third New International Dictionary of the English Language* 11 (2002).

59. *Black's Law Dictionary* 15 (7th ed. 1999).

60. *American Fire & Casualty Co. v. Broeren Russo Construction, Inc.*, 54 F.Supp.2d 842 (C.D. Ill. 1999); *Bituminous Casualty Corp. v. Gust K. Newberg Construction Co.*, 218 Ill.App.3d 956 (1991).

Chapter 7

An Overview of Mold Law in Massachusetts

David M. Governo, Michaela A. Fanning, and Cynthia J. Stephens

Synopsis

7.1 Introduction

For the better part of a decade, indoor air quality issues have been a focus of the technical, scientific and medical communities. Until recently, however, indoor air quality complaints have not been a major factor in toxic tort litigation. Mold has changed this dramatically. Concern over the presence and potential effects of mold has erupted into a scramble of activity on many fronts: sampling, testing, analysis, remediation, potential health effects, and responsibility. As parties, aggrieved by personal injuries and property damage, begin to file suit, litigators scramble to become familiar with the unique issues posed by mold claims. This chapter explores some of the challenges mold claims have presented and describes the current status of the law in Massachusetts.

In Massachusetts, common law causes of action for mold-related damages can be brought under theories of negligence, breach of contract, breach of warranty, and nuisance. Statutory claims can be brought pursuant to landlord-tenant and consumer protection laws. Whether prosecuting or defending a mold case, various procedural issues must be considered as set forth in Section 7.4.

7.2 Massachusetts Common Law Theories

The most common theories asserted by claimants for mold-related injuries are negligence, breach of contract, and breach of warranty.

A. Negligence

A claimant may assert claims for negligence, negligent trespass, and negligent misrepresentation as a result of excessive mold accumulation. Potential defendants include anyone who designs, owns, builds, renovates, repairs, or cleans buildings as well as real estate brokers and managers.

In a negligence action, a plaintiff must prove the existence of a duty owed by the defendant to the plaintiff, a breach of that duty, and actual harm proximately resulting from that breach of duty. *Bacon v. Federal Kemper Life Assurance Co.*, 400 Mass. 850, 853, 512 N.E.2d 941, 943 (1987). Causes of action relating to mold present substantial hurdles given the difficulty of proving a causal relationship between the alleged tortious conduct and mold-related injuries.

Whether a defendant owes a plaintiff a legal duty turns on the existence of a special legal relationship, and the foreseeability of the resulting harm if the duty owed is breached. *Coughlin v. Titus & Bean Graphics, Inc.*, 54 Mass. App. Ct. 633, 639, 767 N.E.2d 106, 111 (2002) quoting *Irwin v. Ware*, 392 Mass. 745, 756, 467 N.E.2d 1292, 1300 (1984). A plaintiff need not show the foreseeability of the particular harm, only harm of the same general character as the harm allegedly suffered. *Jorgensen v. Massachusetts Port Authority*, 905 F.2d 515, 523 (1st Cir. 1990) citing *In Re Sponatski*, 220 Mass. 526, 530-31, 108 N.E. 466-68 (1915).

In the instance of mold, the foreseeability question arises in the context of the relationships between landlords and tenants, as well as contractors and property owners. In the latter instance, for example, a plaintiff must demonstrate it was foreseeable that the design, construction or maintenance of a building would give rise to an accumulation of mold. Since mold-based claims are a relatively new phenomenon, a plaintiff may have difficulty proving that harm of this nature was foreseeable.

If the plaintiff does prove the existence of a duty, he or she must next demonstrate a breach of that duty. The standard used to determine a breach is whether the defendant employed the same level of care a reasonably prudent person skilled in the same trade would exercise under the same or similar circumstances. *Klein v. Catalano*, 386 Mass. 701, 718, 437 N.E.2d 514, 525 (1982). Given the lack of established industry standards regarding the prevention of mold growth, however, it is difficult to predict standards against which the allegedly tortious conduct will be measured. *See Center for Disease Control, State of the Science on Molds and Human Health: Statement of Stephen C. Redd, M.D.*, http://www.cdc.gov/nceh/airpollution/mold at 8. Dr. Redd, a CDC scientist on air pollution and respiratory health, has stated, for example, "there are no quantitative standards, guidelines, or uniform recommendations for responding to mold in indoor environments ...". *Id.* Proving a deviation from accepted practices will require testimony from one or more expert witnesses depending upon the complexity of the defect or conduct which gave rise to the accumulation of mold.

Medical, environmental, and other expert testimony is also essential for a plaintiff to demonstrate the existence of a causal relationship between the defendant's conduct and the plaintiff's injuries or property damage. The proximate cause test requires a showing that (1) the defendant's conduct was a "but-for" cause of the injury, and (2) the defendant's conduct was a "substantial contributing legal factor" in bringing about the plaintiff's harm. *Jorgensen*, 905 F.2d at 524 citing *Tritsch v. Boston Edison Co.*, 363 Mass. 179, 182, 293 N.E.2d 264, 267 (1973); *Wallace v. Ludwig*, 292 Mass. 251, 254, 198 N.E. 159, 161 (1935).

In a claim for property damage, proving a connection between the defect (e.g., water infiltration) and resulting mold may not be all that difficult. The same cannot be said, however, in the personal injury context where the proof of causation is more complex.

Personal injury mold claims are similar to cases involving claims for lead poisoning in that, in both instances, the plaintiff must meet a two-tier proximate cause standard. That is, a plaintiff must show (1) the mold in the plaintiff's body was caused by the offending conduct, structure or defect, and (2) the plaintiff's mold exposure was the cause of the personal injuries alleged. The first tier for proving proximate cause is challenging because mold is ubiquitous and exists naturally in the environment. It is difficult, and perhaps impossible, in some instances, to identify all possible sources of mold which may have contributed to any mold present in the claimant's body. Mold identified in a person could be linked to mold in the home, business, school, soil, water, and air. A serious question also arises as to whether mold which is contained (e.g., behind walls) can even be deemed an actionable source of mold because it was not directly accessible to the claimant.

The second part of the equation, whether the mold to which the person was exposed caused the injuries alleged, poses just as challenging a question. Symptoms which claimants relate to "toxic" mold exposure are typically non-specific and subjective in nature. Potential health effects stemming from "toxic" mold exposure are not well known or documented medically or scientifically. The debate in these scientific fields continues, and there are few medical or scientific studies or other authority upon which a claimant can rely to prove causation. While there is evidence that mold can cause an allergic reaction, at present, there is no definitive link between mold and a toxic, non-allergic human response. *See Center for Disease Control, State of the Science on Molds and Human Health: Statement of Stephen C. Redd, M.D.*, http://www.cdc.gov/nceh/airpollution/mold at 3, 10. There is also insufficient evidence to link an accumulation of mold with the onset of asthma. *Id.* at 3.

The pioneer mold trial in Massachusetts, which addressed the admissibility of expert testimony as well as causation issues, was recently litigated in Essex County. *Stevens v. The Pirates Lane Condominium Trust*, No. 96-00403 (Mass. Super., filed February 27, 1996). In Stevens, the court denied the defendant's motion, *in limine*, to exclude the testimony of the plaintiff's mold expert, Eckardt Johanning, M.D., on the grounds that his anticipated testimony did not meet the standard set forth in *Daubert v. Merrell Dow Pharmaceuticals, Inc.*, 509 U.S. 579 (1993). *See, Stevens* memorandum and order by Judge Agnes, Jr. (June 19, 2002). Dr. Johanning testified at the trial as both a mold expert and as the plaintiff's treating physician. He opined that mold causes infections, allergies, and toxic reactions in humans. He testified that these three categories of effects of molds on humans are generally accepted by the medical community. He was not allowed to testify regarding the plaintiff's specific toxic encephalopathy. Dr. Johanning was permitted to testify, however, that the plaintiff's six-week exposure to mold in the subject premises was of sufficient duration for her to develop a sensitivity to mold.

Whether a medical expert's anticipated testimony satisfied the Daubert standard was also recently considered by the Barnstable Superior Court in the case of *Liska v. Travelers Property Casualty Corp., et al.* No. 99-00569, 2004 Mass. Super. LEXIS 37 (January 28, 2004). *Liska* was a declaratory judgment action involving insurance coverage in a property damage case. Ms. Liska sued her homeowners' insurers alleging her home was unlivable due to mold. She did not claim any personal injury. Since there were two possible sources of the mold in her home's ductwork, Ms. Liska had to prove the mold was caused by a covered event in order to prevail against her insurers. *Id.* at 1.

Prior to trial, a three-day evidentiary hearing on *motions in limine* was held, and the court determined that plaintiff's experts, Jeffrey Adams, Ph.D. and Eckardt Johanning, M.D., were qualified to render causation opinions vis-à-vis mold. *Id.* at 9. The defendants argued that Dr. Johanning's opinion was based upon an incomplete medical history since he never reviewed Ms. Liska's medical records. The defendants also argued Ms. Liska's work environment was contaminated with mold. *Id.* at 8. The court noted the arguments raised by the defendants were best left to rigorous cross-examination of Dr. Johanning at trial. *Id.* at 9. The court held that Dr. Johanning could testify that Ms. Liska was exposed to mold that led to an allergy hypersensitivity reaction. More important to the specific issue in this case, however, he was also allowed to testify about the timing of the commencement of her exposure to the mold. *Id.*

The case was tried, jury-waived, in September 2004. *Liska v. Travelers Property Casualty Corp., et al.*, 99-00569, 2004 Mass. Super. LEXIS 37 (November 19, 2004). The court ruled in favor of both insurance carriers because the mold in the ductwork of the house was not caused by a leaking hot water tank, but by a series of previous water and moisture events that were not covered losses under the policy. *Id.* at 22. The court noted that Dr. Johanning assumed that her symptoms began when the hot water tank leaked. *Id.* at 12-13. Plaintiff's own primary-care physician, however, testified that her respiratory and allergy problems began long before the event involving the leaking hot water tank. *Id.* at 11-12.

In its post-trial opinion, the court noted a difference between a loss that is caused by an excluded event, and a loss caused by a covered event that gives rise to damage, such as mold, which is specifically excluded under the policy. *Id.* at 19 citing Hanover New England v. Smith, 35 Mass. App. Ct. 417, 420-421 and 621 N.E.2d 382, 383 (1993) *(citations omitted)*. The court noted that the failure of the hot water tank (and any resulting mold) was a covered event under the policy but that the leaking water tank was not the proximate cause of the mycotoxins in the ducts. *Id.* at 21. As a result, the plaintiff was unable to prevail against her insurers.

Another Massachusetts (Norfolk County) superior court judge recently afforded considerable deference to the opinions of Dr. Christine Oliver in an action brought to close a courthouse in Dedham, Massachusetts. *Schmidt v. County of Norfolk*, No. 02-00614, 2002 Mass. Super. LEXIS 327 (August 15, 2002).[1] In *Schmidt*, the plaintiffs sought to have the courthouse closed on the grounds that it posed significant health risks to all those who entered the building. *Id.* at 3. One of the many problems was mold accumulation due to inadequate ventilation throughout the building. *Id.* at 8-9. The court noted Dr. Oliver's impressive credentials and experience in the field of occupational medicine. *Id.* at 14-15. She conducted an epidemiologic study among the courthouse employees, and those who spent any time at the courthouse (attorneys, clients, etc.), regarding their medical symptoms, and the extent to which the symptoms abated while they were away from the building. *Id.* at 15-16. The court issued a permanent injunction closing the courthouse based, in large measure, upon the testimony of Dr. Oliver. *Id.* at 44.

Defense counsel must consider using expert testimony to rebut proximate cause testimony by plaintiff's expert and, if applicable, to mitigate damages. Defense experts should stand prepared to discuss the duration of exposure and the level of toxicity of the mold to which the plaintiff was allegedly exposed. Other factors include pre-existing medical conditions, tobacco smoking, and other significant environmental exposures.

As to the question of damages, a plaintiff must demonstrate identifiable physical injuries which were caused or aggravated by exposure to mold. A plaintiff may only be awarded compensatory damages for medical expenses, lost earnings, lost earning capacity, pain and suffering, emotional distress, and loss of consortium once he or she demonstrates that mold brought about the particular damages alleged.

1. Claims of negligence

A claimant may also proceed under the theory of negligent trespass when, for example, the claim involves the flow of water onto property resulting in mold growth. *DeSanctis v. Lynn Water & Sewer Commission*, 423 Mass. 112, 118, 666 N.E.2d 1292, 1296 (1996). To succeed, the plaintiff must demonstrate that the trespassing water caused the mold growth on the subject property. *Id.; Espahbodi v. Town of Sudbury*, No. 94-4230, 1998 Mass. Super. LEXIS 436, at *25-*26 (August 4, 1998).

A plaintiff may also assert a claim for negligent misrepresentation. To prove misrepresentation, the plaintiff must show the defendant made a false statement of a material fact made to induce the plaintiff to act, and that the plaintiff relied on the false statement to his or her detriment. *Ravosa v. Zais*, 40 Mass. App. Ct. 47, 52, 661 N.E.2d 111, 116 (1996), *rev. denied*, 422 Mass. 1108, 664 N.E.2d 1198 (1996) (citations omitted). An action may lie even though the defendant did not know the falsity of the statement when made, if the truth of the matter was reasonably susceptible of actual knowledge. *Acushnet Federal Credit Union v. Roderick*, 26 Mass. App. Ct. 604, 605, 530 N.E.2d 1243, 1244 (1988), citing *Yorke v. Taylor* 332 Mass. 368, 374, 124 N.E.2d. 912, 916 (1955).[2]

2. Statutory and regulatory violations

In Massachusetts, a violation of a statute or regulation does not constitute negligence per se, rather, such a violation is evidence of negligence. St. Germaine v. Pendergast, 411 Mass. 615, 620, 584 N.E.2d 611, 614 (1992).

B. Breach of Contract

Breach of contract claims are commenced in cases between contractors and property owners, or in the landlord-tenant situation. The terms and conditions of the contract or lease determine the nature of the breach. One may also add other parties to the case depending upon the language of the contract. Indemnity provisions may also be included.

Contractual provisions limiting liability are enforceable in Massachusetts. *Canal Electric Co. v. Westinghouse Electric Co.*, 973 F.2d 988, 990 (1st Cir. 1992). Whether a provision will be upheld is determined on a case-by-case basis interpreting the contractual language employed. *Id.* at 992.

C. Breach of Express Warranty

Express warranties in contracts have been upheld by the courts in Massachusetts. *Albrecht v. Clifford*, 436 Mass. 706, 708, 767 N.E.2d 42, 44 (2002). The standard of performance is determined by the promise allegedly made by defendant, rather than a standard set forth by law. *Anthony's Pier Four, Inc. v. Crandall Dry Dock Engineers, Inc.*, 396 Mass. 818, 822, 489 N.E.2d 172, 175 (1986).

Of note, in the *Albrecht* decision, is the fact that the court strictly interpreted the express warranty to defeat the property owner's claim against the contractor. *Albrecht*, 436 Mass. at 718, 767 N.E.2d at 51. The building contract guaranteed the fitness, *inter alia*, of the chimneys of defendant's new home for one year. *Albrecht*, 436 Mass. at 708, 767 N.E.2d at 44. Several years thereafter, the owners learned of the defective installation of the chimneys. *Id.* The court ruled against the owners, noting that their failure even to light the chimneys during the one-year express guaranty period was unreasonable. *Id.* at 715, 767 N.E.2d at 49-50. The court ruled the defective condition of the chimneys was "knowable" to the owners, that is, patent, and dismissed their action as time barred. *Id.*

In *Anthony's Pier Four*, the court recognized the potential for an oral express warranty beyond the actual language of the building contact. *Anthony's Pier Four*, 396 Mass. at 828, 489 N.E.2d at 178. The defendant designed and installed a dock to which the plaintiff moored his yacht/restaurant in 1968. *Id.* at 820, 489 N.E.2d at 174. The ship sank in 1978. *Id.* The court determined that the plaintiff's claim for breach of express warranty did not accrue until the boat actually sank because that was the first time the plaintiff had notice of the defect. *Id.* at 825-26, 489 N.E.2d at 177.

The plaintiff swore that the defendant made an express warranty when he orally assured the plaintiff that extra pilings were not necessary because the existing plans were sufficient to hold the ship. *Id.* at 827, 489 N.E.2d at 177. The defendant denied making such guarantees. *Id.* The court denied the defendant's motion for summary judgment and remanded the case to determine it on its merits. *Id.* at 829, 489 N.E.2d at 179.

D. Breach of Implied Warranty

1. Warranty of Fitness

A person who enters into a contract to do work or to perform a service impliedly agrees to complete the job in a workmanlike manner and to use the skills a reasonably prudent person skilled in such work would exercise under the same or similar circumstances. *Klein*, 386 Mass. at 718, 437 N.E.2d at 525. This theory has obvious implications in a mold case, however, a plaintiff must still prove it was the unworkmanlike conduct which caused the accumulation of mold.

2. Warranty of Habitability

The Massachusetts Supreme Court only recently recognized a cause of action for breach of warranty of habitability in the sale of new homes. *Albrecht*, 436 Mass. at 710-11, 767 N.E.2d at 46-47. The purpose of the law is to protect a consumer from latent defects. *Id.* at 711, 767 N.E.2d at 47. This warranty cannot be waived or disclaimed by the builder. *Id.*

To succeed on this claim, the plaintiff must demonstrate that: (1) he or she purchased the house new from the defendant; (2) the house contained a latent defect; (3) the defect manifested only after purchase; (4) the defect was caused by the builder's improper design, material, or workmanship; and (5) the defect created a substantial question of safety or made the house unfit for human habitation. *Id.* at 711-12, 767 N.E.2d at 47.

A cause of action under this theory is ideal for a new homeowner who has a claim for mold growth. Whether the warranty runs with the new home was not addressed by the court, so it remains unclear whether a subsequent purchaser may maintain a claim against the builder within the statute of limitations. *Id.* at 711, n. 9, 767 N.E.2d at 47, n. 9.

A warranty of habitability is not implied, however, in the transfer of property from one owner to the next. *Solomon v. Birger*, 19 Mass. App. Ct. 634, 638, n. 7, 477 N.E.2d 137, 141 n. 7 (1985), *rev. denied*, 395 Mass. 1103, 481 N.E.2d 198 (1985).

E. Nuisance and Negligent Maintenance

A claimant can also commence an action for a private nuisance for mold under certain circumstances. A private nuisance occurs when an owner of property creates, permits or maintains a condition or activity on property which creates a substantive or unreasonable interference with another owner's use and enjoyment of his property. *Murphy v. Town of Chatham*, 41 Mass. App. Ct. 821, 823, 676 N.E.2d 473, 475 (1996) (citations omitted). One who allows a nuisance to continue may be as liable as the person who created the initial nuisance. *Id.* (citations omitted).

Negligent maintenance of property which, for example, brings about an accumulation of mold is a separate claim from a private nuisance. *Id.* at 825, 676 N.E.2d at 476. The distinction may be relevant if a governmental body is a potential defendant. The Massachusetts Tort Claims Act, ch. 258, § 4, requires a claimant to present, in writing, his or her claim within two years of the occurrence. A failure to meet this statutory requisite is fatal to proceeding with a tort action. The presentment requirement, however, does not apply to claims for a private nuisance. *Murphy*, 41 Mass. App. Ct. at 825, 676 N.E.2d at 476-77. A claimant may proceed with a private nuisance claim against the government, commenced more than two years after the occurrence, subject, of course, to the three-year statute of limitations. *Id.* at 824, 676 N.E.2d at 476. For an example of a nuisance and negligent maintenance mold case, *see Espahbodi v. Town of Sudbury*, No. 94-4230, 1998 Mass. Super. LEXIS 436, discussed in Section 7.5.

7.3 Massachusetts Statutory Claims

The landlord-tenant and consumer protection statutes have wide application and significant consequences in mold cases, the most important being the ability to recover attorneys' fees, costs, and sometimes, punitive damages.

A. Landlord-Tenant Laws

1. Quiet Enjoyment

Massachusetts General Laws, ch. 186, § 14 codifies a tenant's common law rights to the quiet enjoyment of the premises. *Al-Ziab v. Mourgis*, 424 Mass. 847, 850, 679 N.E.2d. 528, 529 (1997). A finding of liability under this statute requires a showing of negligence by the landlord. *Id.* at 850, 679 N.E.2d at 530. Violation of the sanitary code, alone, is not sufficient, since the landlord must have notice of the violation. *Id.* at 850-51,

679 N.E.2d at 529-30. Under ch. 186, § 14, the landlord can be liable for actual and consequential damages, or three months' rent, whichever is greater. The case of *Curtis v. Surrette*, 49 Mass. App. Ct. 99, 104, 726 N.E.2d 967, 971 (2000), sets forth a good example of how to calculate damages and fees under the quiet enjoyment statute.

What gives this statute and the reprisal and consumer protection law discussed below their impact is the ability to recover attorneys' fees and costs. Even a small damage claim, resulting in a minimum recovery of three months rent, can be greatly enhanced by an award of fees and costs.

2. Warranty of Habitability

Another Massachusetts landlord-tenant statute, the warranty of habitability, encompasses both common and statutory law. The landmark case of *Boston Housing Authority v. Hemingway*, 363 Mass. 184, 199, 293 N.E.2d 831, 843 (1973), held that an implied warranty of habitability existed in a lease of residential properties. Every landlord warrants to its tenants that the premises will be delivered and maintained in a habitable condition. Id. at 199, 293 N.E.2d at 843. In *Crowell v. McCaffrey*, 377 Mass. 443, 447, 386 N.E.2d 1256, 1259 (1979), the court extended the warranty to encompass liability for personal injuries caused by a breach. *See also Simon v. Solomon*, 385 Mass. 91, 96, 431 N.E.2d 556, 561(1982).

Separate from the common law is the statutory version of the warranty of habitability, M. G. L. ch. 239, § 8A. Proof that the premises is in violation of the state sanitary code creates a presumption that entitles the tenant to a defense against an eviction for the non-payment of rent. Tenants can withhold rents or initiate sanitary code enforcement proceedings against landlords who refused to remedy code violations. *Hemingway*, 363 Mass. at 199, 293 N.E.2d at 843. If the tenant is entitled to relief, he can offset his rent, or recover the difference of the value of the apartment as warranted and the rental value in its defective condition. *Id.* at 201-02, 293 N.E.2d at 844-45.

Attorneys' fees and costs, however, are not available under the warranty of habitability. Additionally, duplicative recoveries under the quiet enjoyment statute and warranty of habitability statute are also not allowed. *Wolfberg v. Hunter*, 385 Mass. 390, 401, 432 N.E.2d 467, 474 (1982); *Darmetko v. Boston Housing Authority*, 378 Mass. 758, 761-62, 393 N.E.2d 395, 398 (1979). Since the remedy for a breach of the warranty of habitability is not as extensive as the remedy under the quiet enjoyment statute, the habitability claim may not be of great value to a plaintiff. When used as an alternative in the event that negligence actions are susceptible to summary judgment, however, the warranty of habitability has significant consequences and should not be ignored.

3. Reprisal Statutes

There are also Massachusetts landlord-tenant statutes that provide relief for a tenant when a landlord retaliates against him or her for reporting code violations, such as excessive mold growth.

Pursuant to M. G. L. ch. 186, § 18, any landlord who threatens a tenant for reporting violations is liable for one to three months' rent or actual damages, whichever is greater, including costs and attorneys' fees. *Manzaro v. McCann*, 401 Mass. 880, 883, n. 3, 519 N.E.2d 1337, 1340, n. 3 (1988). Additionally, ch. 239, § 2A creates a rebuttable presumption that an eviction is a reprisal against a tenant who has reported a code violation. *Appelstein v. Quinn*, 361 Mass. 861, 281 N.E.2d 228, 229 (1972). A landlord must take great care, therefore, before initiating eviction actions against a tenant who had reported mold accumulation in the premises.

B. Massachusetts Consumer Protection Law

The Massachusetts Consumer Protection Law, Chapter 93A, is a far-reaching statute which prohibits unfair acts in the context of trade or commerce. Under section 9 of the statute, if there is a willful violation of Chapter 93A, double or triple damages can be awarded. This is the third Massachusetts statute that allows for the recovery of attorneys' fees and costs. Unlike the landlord-tenant statutes, however, Chapter 93A applies to anyone who is involved in a business transaction. It typically involves consumer transactions, such as those between a homeowner and a contractor, but can also include business to business transactions. A demand letter is a prerequisite to a lawsuit under section 9 of the statute. *Cassano v. Gogos*, 20 Mass. App. Ct. 348, 350, 480 N.E.2d 649, 651 (1985); *Spring v. Geriatric Authority of Holyoke*, 394 Mass. 274, 289, 475 N.E.2d 727, 736 (1985).

This statute is particularly important in mold cases since it potentially increases the typical recovery found in a tort action. Not only can a plaintiff recover attorneys' fees and costs under this law, he or she may also be entitled to multiple damages. In the personal injury context, attorneys' fees and costs can be extensive given the need for one or more expert opinions to succeed on the claims alleged. When punitive damages are awarded, recovery can be significant. The case of *Brossi v. Fisher*, 1999 Mass. App. Div. 99 (1999) sets forth an analysis of Chapter 93A liability and calculation of damages in a landlord-tenant dispute where damages included mold accumulation.

Chapter 93A claims also encompass an insured's claim against his insurance company for "bad faith" in denying coverage or failing to respond timely to a mold infiltration claim. *See also* M.G. L. ch. 176D, § 3 for a description of unfair methods of competition and unfair and deceptive acts and practices in the insurance business.

7.4 Procedural Considerations

A. Statutes of Limitation

1. Negligence

Pursuant to M. G. L. ch. 260, § 2A, a person filing a claim for personal injuries must bring the claim three years from the date the claimant discovers the personal injury or property damage, and its cause. Under section 7 of the statute, a minor plaintiff has until three years from the date of his or her eighteenth birthday to file suit. These statutes begin to run when the claimant discovers or, in the exercise of reasonable diligence, should have discovered the harm. *Riley v. Presnell*, 409 Mass. 239, 244, 565 N.E.2d 780, 785 (1991).

An exception to the three-year tort action limitation may exist if the tortious conduct is deemed continuous. *Doe v. Blandford*, 402 Mass. 831, 839, 525 N.E.2d 403, 409 (1988) (citations omitted). In a wrongful death case, a Massachusetts (Middlesex County) superior court judge recently addressed whether repeated flooding, resulting in the growth of mold over a seven-year period, constituted a "continuous tort" for purposes of the statute of limitations. *Freeman v. Massachusetts Institute of Technology*, No. 00-0204, 2000 Mass. Super. LEXIS 311, at *7-*11 (August 14, 2000). In Freeman, the plaintiff alleged that her apartment initially flooded in 1990. *Id.* at *2. Problems with flooding and the ensuing growth of mold, bacteria, and mildew continued through 1996. *Id.* In September of 1996, the city ordered the owner of the property to correct the cause of the flooding. *Id.*

In 1991, the plaintiff's son was diagnosed with severe bronchial asthma, with a strong allergic component. *Id.* His exposure to mold, and other allergens related to the flooding caused chronic inflammation of his lungs. *Id.* at *2. Two months after the September, 1996 flood, the child was hospitalized with a collapsed lung following a severe asthma attack. *Id.* at *3. The child subsequently died on July 30, 1997. *Id.*

The plaintiff presented her claim to the city on July 7, 1999. *Id.* In her presentment letter, the plaintiff claimed that the family only "recently" learned that the mold problem in the apartment had exacerbated her son's asthma. *Id.* at *3-*4. The plaintiff filed her complaint on January 14, 2000. *Id.* at *4. The city filed a motion for summary judgment in which it argued that all claims relating back further than January 15, 1997 were time barred.[3]

The court stated that a continuous tort does not occur where conduct occurring prior to the statute of limitations continues to cause adverse effects or damages beyond the three-year limitation. Id. at *5-*7. A continuous tort only tolls the statute of limitation where the *tortious act continues, not where the continuing effects of the initial tort continue. Id.*, citing *Maslauskas v. United States*, 583 F. Supp. 349, 351 (D. Mass. 1984). *See also Cooper v. United States*, 442 F.2d 908, 911-12 (1971) (claim time barred because of the government's refusal to provide medical care was not continuous despite the claimant's repeated requests for medical attention). In *Cooper*, the court determined the cause of action accrued when it became apparent to the claimant that no medical treatment was forthcoming. *Id.* at 912.

The *Freeman* court went on to say that, if the problem with flooding was caused by the negligent design and installation of the sewer system, the conduct was not continuous throughout the tenancy, and the claims would be time barred. Freeman, 2000 Mass. Super. LEXIS 311, at *9-*10. If, on the other hand, the city's tortious conduct was in the nature of failing repeatedly to clear blockages from the sewers, such conduct could constitute a continuing tort, and these claims could be deemed timely. *Id.* at **11*.

2. Breach of Contract

Under M. G. L. Ch. 260, § 2, a claim for breach of contract must be filed within six years after the cause of action accrues.

3. Breach of Express Warranty

The claimant must assert a claim for breach of express warranty under a breach of contract theory, and therefore, has six years from the accrual of the cause of action within which to file the claim. *Anthony's Pier Four*, 396 Mass. at 822, 489 N.E.2d at 174. Since the action does not sound in tort, the statute of repose does not apply. *Id.* at 823, 489 N.E.2d at 175.

4. Breach of Implied Warranty

A claim for breach of an implied warranty is considered a tort which must be commenced within the three-year statute of limitations and the six-year statute of repose. *Albrecht*, 436 Mass. at 712, 767 N.E.2d at 47.

5. Violation of M. G. L. ch. 93A

A claimant has four years within which to file a claim for breach of the consumer bad-faith statute. A claimant may still pursue a claim under Ch. 93A even if the underlying tort action is barred by the three-year statute of limitations (assuming the ch. 93A claim is brought within the four-year time limitation). *Fine v. Huygens et al.*, 57 Mass. App. Ct. 397, 401, 783 N.E.2d 842, 846 (2003).

B. Claims Against The Government

Massachusetts General Laws Ch. 258 provides procedural limitations for claims brought against the Commonwealth of Massachusetts, its municipalities, counties, districts, and the officers and employees thereof. Under section 4 of the statute, a claimant must present, in writing, his or her claim within two years after the date upon which the cause of action arose. Failure to make a proper presentment of a claim could result in the dismissal of the claim by the court. *Doe*, 402 Mass. at 839, 525 N.E.2d at 408.

Note that, in a claim for nuisance, the claimant need not present the claim under this statute. *Murphy*, 41 Mass. App. Ct. at 824-25, 676 N.E.2d at 476-77. According to section 10 of the statute, presentment is also not required for intentional torts and other limited actions by the government.

C. Statute of Repose

The Massachusetts statute of repose creates a second time limitation and is designed to protect those engaged in the construction industry. Massachusetts General Laws Ch. 260, § 2B sets forth time limitations in the context of torts arising from "improvements to real property." A tort claim for damages relating to "deficiency or neglect in the design, planning, construction or general administration of an improvement to real property" (not including a public agency) must be brought within three years after the cause of action accrues, however, such an action cannot be commenced more than six years after the earlier of the dates of:

- the opening of the improvement to use; or
- substantial completion of the improvement and the taking of possession for occupancy by the owner.

The same three-year limitation applies to a public agency, however, the six-year limitation applies to the earlier of the dates of:

- official acceptance of the project by the public agency;
- the opening of the real property to public use;
- the acceptance by the contractor of a final estimate prepared by the public agency pursuant to M. G. L. ch. 30, § 39G; or
- substantial completion of the work and the taking possession for occupancy by the awarding authority.

The term "improvement" has been defined as "a permanent addition to or betterment of real property which enhances its capital value, and that involves the expenditure or labor or money and is designed to make the property more useful or valuable as distinguished from ordinary repairs." *Snow v. Harnischfeger Corporation*, 12 F.3d 1154, 1161 (1st Cir. 1993).

Parties protected by this statute include, *inter alia*, architects, engineers, and contactors. The statute may also bar claims against a manufacturer or supplier if the product is produced for a particular project and meets the specifications of an architect or engineer. *Id.* at 1160; *Rosario v. M.D. Knowlton Co.*, 54 Mass. App. Ct. 796, 801, 767 N.E.2d 1126, 1131 (2002). The court in *Snow* stated that the statute of repose defines protected actors by reference to the acts protected under the statute. *Snow*, 12 F.3d at 1160; *See also Dighton v. Federal Pacific Electric Co.*, 399 Mass. 687, 695, 506 N.E.2d 509, 515 (1987) (citations omitted). The Dighton court determined that a circuit switch manufacturer was not a protected party under the statute even though the circuits became a permanent part of the structure. *Id.* at 699, 506 N.E.2d at 517. This case is also significant because the court ruled that a claim for contribution could not stand against an architect because the statute of repose barred the plaintiff's claims against the architect. *Id.* at 692-93, 506 N.E.2d at 513.

The relation-back doctrine, set forth in Mass. R. Civ. P. 15(C), does not apply in cases in which the statute of repose applies. *Tindol v. Boston Housing Authority*, 396 Mass. 515, 518, 487 N.E.2d 488, 490 (1986), *Fine v. Huygens,* 57 Mass. App. Ct. 397, 401, 783 N.E.2d 842, 846 (2003), and *Ferrera & Sons, Inc. v. Samuels & Hannan Construction*, 21 Mass. App. Ct. 170, 172-173, 486 N.E.2d 58, 60 (1985). In *Ferrera*, for example, the plaintiff was precluded from amending its complaint to add additional parties, and the defendants were barred from bringing third-party claims for indemnification and contribution against the additional defendants after the six-year repose limitation.

It does not matter that the defect was not discovered, nor could it reasonably have been discovered within the six-year deadline under the repose statute. The statute of repose completely eliminates a cause of action after six years. *Ferrera*, 21 Mass. App. Ct. at 173, and 486 N.E.2d at 60. This includes claims for indemnification and contribution which can be brought within one year of a settlement or judgment in a non-construction context.

Unlike the statute of limitations, the statute of repose does not create an extension of time (i.e. three years from the date a minor claimant turns eighteen) within which a claim may be brought by a minor. *Tindol v. Boston Housing Authority*, 396 Mass. 515, 518, 487 N.E.2d 488, 490 (1986).

This statute is significant in the context of mold because it could result in a cause of action for mold being barred after six years regardless of whether mold was present in the improvement to real property during the six-year period.

7.5 Overview of Massachusetts Mold Cases

There has been only one mold jury trial in Massachusetts to date. *Stevens v. The Pirates Lane Condominium Trust*, No. 96-00403 (Mass. Super., filed February 27, 1996). The plaintiff claimed she experienced physical symptoms six weeks after moving into her one-bedroom condominium in 1995. She alleged property damage, real estate loss, and physical injuries including asthma and immune-system compromise due to the inhalation of toxic mold. During the eight-day trial, the plaintiff's attorney argued the defendant condominium trust knew the plaintiff's garden-level unit had a chronic wetness problem, but failed to remedy the water seepage into the condominium. In June of 2002, the jury returned a verdict for the plaintiff in the amount of $285,000 to which $264,326 in interest was added.

In addition to the cases that have been discussed herein, there is one case which highlights the challenges a plaintiff faces in proving causation as well as damages. In *Espahbodi*, No. 94-4230, 1998 Mass. Super. LEXIS 436, at *1, the plaintiff homeowner's property was bisected by a drainage easement held by the town. Over the years, the house became increasingly musty and humid. *Id.* at *14. The wallpaper peeled, the kitchen floor tile cracked, the garage floor was damp, the basement was chronically wet, moss grew on two sides of the house, and wetlands were encroaching on the property's backyard. *Id.* at *14-*16. A tenant in the property complained of excessive mold accumulation, and a state sanitary code violation notice was sent to the plaintiff. *Id.* at *19. The homeowner sued the town for nuisance, trespass, negligence, and diminution of the value of their property as the result of the town's negligent maintenance of the easement. *Id.* at *1. The court held that while there was evidence that the easement was negligently maintained, there was no evidence of a causal connection between the negligence and the moisture in the house, and further, there was not enough evidence to prove a diminution in value of the house, regardless of the cause. *Id.* at *21-*26.

7.6 Insurance Considerations

The proliferation of mold claims is of great concern to the insurance industry. Several Massachusetts insurers have adopted limitations for mold in their homeowners' policies which exclude coverage for mold unless the condition resulted from a "covered peril", such as water damage from an accidental discharge or overflow of water, or windstorm. Some policies also include an endorsement with a separate limit for mold. *Abacus Insurance Brokers Insurance Products, Questions and Answers*, http://www.abacusins.com/products/qa.aspx?qaid=140. Insurance coverage is clearly important to both plaintiffs and defendants, but plaintiffs especially need to consider this issue when deciding whether to sue and whom to sue.

Endnotes

1. There is no mention in the opinion whether the defendant challenged Dr. Oliver's anticipated testimony under *Daubert.*

2. A plaintiff may also allege intentional misrepresentation but has a higher burden of proof than in the case of negligent misrepresentation. The plaintiff would have to show that the defendant made the claim knowing it was false for the purpose of inducing the plaintiff to act. *Barrett Associates v. Aronson,* 346 Mass. 150, 152, 190 N.E.2d 867, 868 (1963).

3. The plaintiff was statutorily required to present her claim to the city within two years of the occurrence, hence, the defendant argued all claims prior to the two-year period of the presentment letter should be dismissed. *Id.* at *5.

Chapter 8

An Overview of Mold Litigation in New Jersey

Patrick J. Perrone, Esq., and Richard J. Valladares, Esq.

Synopsis

8.1 Introduction

The optimistic assumption that mold litigation was an anomaly confined to a few geographic areas with continual warm, humid weather has all but been abandoned in light of the recent surge of mold-related claims across the nation. Mold spores are present everywhere, and although mold growth flourishes in locations with warm, humid weather, there are various other conditions that may cause indoor mold such as building envelope construction defects.

In addition, mold-related liability claims are no less contingent on specific weather conditions than they are on heightened publicity, rapid demand for new housing, and mounting interest from the plaintiffs' bar. The effect of these unconventional factors is readily apparent in New Jersey. New Jersey's population growth of approximately 78,000 people each year and the consequential demand for new housing and development combined with recent media hype has created an acute awareness of mold presence and mold-related personal injury and property damage allegations within New Jersey's general population. This acute awareness has resulted in a proliferation of mold-related claims in New Jersey. Although the epicenter of such claims is in the insurance arena, the unfortunate fallout from the recent appearance of mold exclusions in insurance policies and the denial of insurance coverage for these claims, is likely to be more mold litigation against builders and property owners. Mold litigation is made even more likely by the recent nation-wide focus by plaintiffs' attorneys on recent allegations that serious bodily injury is caused from mold exposure.

Litigators serving their respective clients should be well-versed on relevant New Jersey legal principles which define applicable causes of action and offer defenses that may lessen the liability blow of a mold lawsuit. Although case law in the mold litigation context is not as well-developed in New Jersey as it is in other states (such as California and Florida) there is no reason to believe these general legal principles will not be wholly applicable to the mold phenomenon.

8.2 Potential Causes of Action

Mold-related lawsuits may encompass a variety of causes of action including, but not limited to, claims founded in negligence, breach of contract, breach of implied warranties, misrepresentation and fraud. These claims spring from the unique responsibilities associated with the construction, sale, operation, management, control, maintenance, and/or renovation of residential or commercial structures. Although these duties and responsibilities have not been fully defined by New Jersey courts in the mold context, the following sections identify the most probable application of New Jersey legal principles to the mold litigation phenomenon.

A. Negligence

As in any negligence action, there are four basic requirements to presenting a *prima facie* case of negligence. The plaintiff must show that the defendant owed a duty of care, that the defendant breached that duty, that the defendant's breach was the actual and proximate cause of the alleged property damage or bodily injury, and that there was actual property damage or bodily injury.[1] Although contractors, subcontractors, architects, building supply manufacturers, maintenance companies, and property owners all have unique duties to a homeowner or tenant, they will often be sued by the same plaintiff in the same case. Exemplifying the broad exposure to a wide array of professionals susceptible to mold litigation is a January, 2003 lawsuit filed in Middlesex County, New Jersey. In that suit, home purchasers simultaneously sued previous homeowners, real estate agents, and home inspectors.[2] Specifically, plaintiffs alleged that the previous homeowners intentionally concealed water problems and other hazardous conditions on the disclosure statement; that the real estate agents knew of, and intentionally misrepresented, the true condition of the home; and that their home inspector provided inaccurate inspection findings.[3]

The typical array of potential defendants in a mold case may be further broadened by virtue of the fact that New Jersey courts have abandoned concepts such as privity[4] and *caveat emptor*.[5] The following sub-sections describe the nature of the duties imposed on a broad range of professionals.

1. Professional sellers of real estate

Historically, real estate purchases were governed by the doctrine of "caveat emptor" or "let the buyer beware." Under this doctrine, the sellers of real estate had no obligation to disclose defects unless otherwise agreed. Currently, state law concerning a seller's duty to disclose varies. Nevertheless, even in those states that follow *caveat emptor* and impose no duty to disclose patent defects, a seller is nonetheless required to disclose known latent defects.[6] Although real estate non-disclosure cases in New Jersey involve more conventional problems such as termite infestation, there appears to be no apparent reason these principles would be inapplicable in a mold case. In fact, mold may be as latent as termite damage, existing behind walls, in basements, or behind drop-ceilings. Moreover, even in cases in which mold conditions have already been remediated, and the sellers might be tempted to keep that remedial information to themselves by arguing that the property is no longer in an allegedly "defective condition," such an option presents high risk to the non-disclosing seller if mold growth re-occurs (which is possible under certain conditions).

Disclosures by a professional seller of residential real estate should be carefully considered in New Jersey, since New Jersey courts reject the *caveat emptor* doctrine entirely, finding that the doctrine is inconsistent with modern home buying practices.[7] New Jersey courts have noted that the principal reasons for imposing a duty to disclose on professional sellers, relates to the difference in bargaining power between the professional seller of real estate and the purchaser of such housing, as well as the difference in access to information between the seller and the buyer.[8] Therefore, the explicit disclosure of discovered mold growth and prior mold remediation should be carefully considered by professional sellers of real property, since such conditions might be deemed a "material" defect by courts or juries.

Pending legislation should also be considered. Currently pending in New Jersey is a toxic mold bill which expressly requires the disclosure of the presence of mold by persons selling or leasing commercial or industrial property, public entities that own, lease, or operate property, lessors of residential property, or lessees who know that mold is present in a property (to inform the owner of the property).[9] Unlike legislation in other states, there is no immunity provision in the New Jersey bill. Thus, although landlords may be required to divulge to prospective tenants the presence of any known mold, and also to provide notice of any prior mold tests and/or remediation, the landlord may nonetheless be liable in any action based on the presence of, or propensity for, mold growth in a building.

2. Builders

Builders are most commonly sued for construction defects causing moisture and water intrusion that fosters mold growth. Plaintiffs typically allege that such defects constitute a breach of the builder's implied duty to carry out its work in a reasonable and prudent manner. Common mold-related defects are those associated with the building envelope, roof, windows, venting, or plumbing. In New Jersey, a builder's duty may extend not only to parties in direct privity, but also to third parties who can establish that the builder's negligence was the proximate cause of their injuries (whether personal injury or property damage).[10]

New Jersey's "New Home Warranty and Builders' Registration Act" specifically protects new homeowners and is likely to apply to certain mold claims.[11] The New Home Warranty and Builders' Registration Act provides that a builder of a new home is liable to any owner thereof for any defect covered by the Act.[12] Warranty periods established by the Act are as follows: (1) one year from and after the warranty date, the dwelling shall be free from defects caused by faulty workmanship and defective materials due to noncompliance with the building standards as approved by the Commissioner of Community Affairs; (2) two years from and after the warranty date, the dwelling shall be free from defects caused by faulty installation of plumbing, electrical, heating and cooling systems, except in the case of appliances in which the manufacturers' warranty governs; and (3) ten years from and after the warranty date, the dwelling shall be free from major construction defects as defined in the Act.[13] Notwithstanding the existence of the Act, homeowners are not limited to submitting a claim under the New Home Warranty and Builders' Registration Act and are statutorily granted the right to pursue any remedy legally available to them.[14]

3. Sub-contractor liability

A recurring question in the construction setting is whether an owner can file a claim against a subcontractor with whom it has no contract.[15] This is a significant issue in mold litigation since water intrusion problems are often identifiable and assignable to individual subcontractors (e.g., window installation, roofing installation, and flashing installation). In most jurisdictions, the lack of privity between an owner and a subcontractor impedes the pursuit of claims directly against subcontractors.[16] In light of this privity dilemma, jurisdictions that recognize an owner's right of recovery against a subcontractor rest liability not on contract principles, but on tort principles.

New Jersey appears to be one of the few jurisdictions that have held that a subcontractor may be held directly liable in tort to an owner or developer even in the absence of a direct contractual relationship. In *Juliano v. Gaston*, a New Jersey court permitted an owner to file a tort action against various subcontractors for failure to carry out their work with reasonable care and skill.[17] In *Gaston*, homeowners brought an action alleging that they had been damaged by the negligent workmanship of two subcontractors involved in their home's construction.[18] The general allegations were that the work of these defendants "was done in a negligent, unworkmanlike manner, contrary to the standards of the industry and in all respects carelessly and improperly" and that "as the direct and proximate result thereof," the home was damaged and had to be repaired (at a significant cost).[19] Although the trial court granted summary judgment to the subcontractors, the New Jersey Appellate Court reversed and remanded.[20] In doing so, the New Jersey Appellate Court rejected the trial court's finding that the action was barred due to lack of privity between the parties, noting that the claim was based on the subcontractors' negligence and had no relation to the contract between the builder and the subcontractors.[21] In so ruling, the court stated:

> We fail, therefore, to understand the concern of the trial judge with the question of privity. Privity is a concept having relevance only in a contractual situation and is, of course, that concept which, as an historical matter and prior to the evolution of the doctrine of strict liability in tort, limited a consumer's warranty recourse to the party with whom he had directly dealt.
>
> . . .
>
> The fact, however, remains that this action by these plaintiffs against these subcontractors was neither instituted nor prosecuted on an implied warranty theory but rather on a traditional negligence basis. Privity is hence irrelevant.[22]

The language of the *Gaston* court's holding revealed an intent to place New Jersey among those jurisdictions that hold a subcontractor directly liable in tort to an owner or developer in the absence of a direct contractual relationship, if that subcontractor's defective workmanship caused the owner to sustain property damage.

Other jurisdictions have consistently so held, recognizing a direct cause of action in negligence, irrespective of the availability of a warranty or strict liability cause of action, when the purchaser of a new house seeks to recover property damages only from a negligent subcontractor.[23]

The significance of this decision should not be overlooked. The *Gaston* decision not only creates a viable cause of action, but seems to permit a plaintiff to elude the consequences of the "economic loss doctrine," which normally prohibits tort actions when the damage is economic only (e.g., the cost to remediate mold growth). In *Gaston*, the "economic loss doctrine" was held to be inapplicable due to the fact that the action was founded on "pure negligence."[24] In so holding, the *Gaston* court stated,

> It appears that the nature of the damages here may be limited to replacement and repair of defective workmanship, a category of damages customarily referred to as "economic loss" .... We see, however, no impediment, conceptual or practical, to the recovery of this category of damages in a negligence action by the purchaser against the subcontractor.
>
> ...
>
> In our view, if plaintiffs sustained any economic loss by reason of the negligence of the subcontractors, they are entitled to be made whole.[25]

B. Breach of contract

Similar to most jurisdictions, a *prima facie* breach of contract case requires the plaintiff to establish: a valid contract, defective performance by the defendant, and resulting damages.[26] Such claims can be expected in the mold context.

C. Breach of implied warranties

In addition to liability for breach of the express terms of a contract, contractors and subcontractors are frequently held liable for breach of the following implied warranties: (1) the implied warranty of good workmanship; (2) the implied warranty of habitability; and (3) the implied warranty of quiet enjoyment.

1. Implied warranty of good workmanship

The most common cases will involve construction defects that lead to water intrusion problems and mold growth. In conjunction with an allegation that such defects are in contravention to the express sales contract, the property owner may also allege that the faulty or substandard construction constitutes a breach of the implied warranty of good workmanship (which applies to all new homes).[27]

The implied warranty of good workmanship is not limited to the initial home purchaser. In New Jersey, subsequent buyers may also be entitled to implied warranty protections.[28]

2. Implied warranty of habitability

In certain mold cases, plaintiffs may allege that the construction defect and resulting mold growth affected the habitability of the property.[29] The implied warranty of habitability means that the property is suitable for living purposes and able to be occupied.[30] Thus, a successful breach of the implied warranty of habitability claim requires proof that some vital living element of the property was affected.[31] Such claims are likely to appear in lawsuits involving allegations of bodily injury (allergies, respiratory problems, fatigue and so on).

3. Implied warranty of quiet enjoyment

Implied warranty of quiet enjoyment claims usually arise in the landlord-tenant context. Under the implied warranty of quiet enjoyment, any act or omission of the landlord which renders the premises substantially unsuitable for the purpose for which they were leased, or which seriously interferes with the beneficial enjoyment of the premises, is a breach of the covenant of quiet enjoyment.[32] The remedy of constructive eviction has been applied to such cases.[33]

Implied warranty of quiet enjoyment claims have often been successfully founded on circumstances regarding health and safety (i.e., claims involving noxious odors[34] or failure to heat the premises[35]). Accordingly, there is no reason to believe that such claims will not be made in the mold context.

D. Fraud or negligent misrepresentation claims

Fraud claims require the plaintiff to prove: (1) a material misrepresentation or omission of a presently existing or past fact; (2) knowledge or belief by the defendant of its falsity; (3) an intention that the other person rely on it; (4) reasonable reliance thereon by the other person; and

(5) resulting damages.[36] Negligent misrepresentation claims require the plaintiff to prove that an "incorrect statement was negligently made and justifiably relied upon" and that the injury was sustained as a consequence of that reliance.[37] When a builder, subcontractor, landlord, property owner, or real estate professional knows of a mold problem and intentionally or negligently fails to advise the tenant or homeowner, a fraud and/or negligent misrepresentation claim is likely to be made.[38]

Beyond these common law duties, the New Jersey Consumer Fraud Act must be factored into disclosure considerations of real estate professionals.

New Jersey Courts have ruled that brokers, agents, and salespersons involved in the sale of real estate are subject to claims brought under the New Jersey Consumer Fraud Act.[39] The Act prohibits both affirmative misrepresentations and acts of omission.[40] The Act states that "the omission of any material fact with intent that others rely upon such . . . omission, in connection with the sale ... of ... real estate" is an "unlawful practice."[41] Such may be the case when the nondisclosures rise to conduct equivalent to "swindling" or "shocking to the ethical sense of the community."[42] Significantly, violation of the Act permits a purchaser of real estate to recover treble damages for any ascertainable loss, plus reasonable attorney's fees.[43]

E. Workers' compensation

Employers and their insurers are also not immune from mold litigation. Notwithstanding the fact that the link between bodily injury and mold contamination has yet to be established, claims for workers' compensation due to injuries related to mold exposure are growing. In mold contaminated buildings, employees will often seek workers' compensation for personal injury allegedly caused by exposure to mold. When mold is identified by an employee, the employer should act accordingly to avoid allegations of "intentional harm," which permits an employee to escape the workers' compensation bar.

New Jersey is no stranger to this phenomenon. In a recent New Jersey case, a workers' compensation judge held that a teacher had suffered 10-percent permanent partial disability after working for several months in a classroom where mold and mildew were present.[44] The judge awarded the teacher $50,682 in compensation for temporary and permanent disabilities and ordered the school board to pay a penalty for denying the teacher's request for a transfer to another school following the onset of her health problems.[45]

8.3 Class Actions

Mold lawsuits will often be brought by numerous similarly situated plaintiffs who are members of a condominium association, tenants in an apartment building, homeowners in a townhouse complex, or home purchasers from a common builder. In such cases, the plaintiffs may seek class certification. To qualify as a class action, a lawsuit must satisfy the four requirements of New Jersey Court Rule 4:32-1: (1) numerosity; (2) commonality; (3) typicality; and (4) adequacy of representation.[46] Although common issues of fact and law might exist with respect to certain limited issues in a mold case, the individual issues are likely to predominate.

8.4 Potential Defenses

A. Statute of repose

A statute of limitations computes the period within which an action must be commenced from the accrual of the cause of action.[47] On the other hand, a statute of repose operates without regard to the accrual of a cause of action.[48] New Jersey has a ten-year statute of repose which is found at N.J.S.A. 2A:14-1.1, and limits the potential liability to which architects and building contractors (among others) are subject.[49] Thus, plaintiffs instituting an action in New Jersey alleging property damage or bodily injury claims based on deficiencies in design, planning, surveying, supervision, or construction of an improvement to real property have only ten years from "substantial completion" of the improvement to real property to file their claim.[50] Significantly, this is so even if the defect is undiscovered until some time after the improvement is made.

Product design alone is not enough to trigger the applicability of N.J.S.A. 2A:14-1.1.51 The statute does not benefit manufacturers and sellers of products who were not involved in the design, planning, and/or construction of the improvement to real estate.[52] More specifically, the sale of a mass produced item does not constitute the "design, planning, supervision or construction of an improvement to real property" as envisioned by the statute.[53] However, when a manufacturer or seller of building products or components significantly participates in the installation of that product into the building, such that it constitutes an "improvement to real property" within the meaning of N.J.S.A. 2A:14-1.1, that party is entitled to protection of New Jersey's statute of repose.[54]

The statute of repose poses serious risks to those representing mold plaintiffs, which is heightened by the fact that mold growth is often a slow process and not discovered until some significant time after the alleged construction defect is identified. For those representing defendants, the statute of repose offers an invaluable defense that summarily defeats a plaintiff's claim.

B. Spoliation

Fulfilling one's duty to preserve evidence is difficult in a mold case. Indeed, both the owner and the tenant will want to quickly remove and remediate any discovered mold. This is especially so in light of the propensity for mold to re-occur when improperly remediated. As part of the remediation process, owners commonly destroy and remove sheet-rock, ceiling tile, insulation, wood framing, or other affected materials. Notwithstanding the reasonableness of such conduct, the destruction of such building components may destroy valuable evidence in a subsequently brought litigation.

Depending on the specific factual circumstances involved, such spoliation of evidence could result in: (1) dismissal of the claim; (2) precluding the offending party from relying on the spoliated evidence; or (3) injecting an "adverse inference" into the jury charge.[55] Thus, owners are placed in the untenable position of having to remove mold laden materials from the property, while at the same time having to "preserve" evidence in the event of subsequent litigation. Under the circumstances, litigators should counsel clients to notify all potentially interested parties prior to demolition.

Spoliation of evidence in a prospective civil action occurs when evidence pertinent to the action is destroyed, thereby interfering with the action's proper administration and disposition.[56] The duty to preserve evidence arises when there is: (1) pending or probable litigation involving the defendants; (2) knowledge by the plaintiff of the existence or likelihood of litigation; (3) foreseeability of harm to the defendants (or in other words, discarding the evidence would be prejudicial to defendants); and (4) evidence relevant to the litigation.[57] The duty to preserve evidence is a question of law to be determined by the court.[58] Although the scope of the duty to preserve evidence is not boundless, an opportunity for inspection should be afforded a potentially responsible party before relevant evidence is destroyed.[59]

Significantly, the spoliator's level of intent, whether negligent or intentional, does not affect the spoliator's liability.[60] Rather, it is simply a factor to be considered when determining the appropriate sanction for the spoliation.[61] Moreover, when a party's agent (i.e., expert witness or legal counsel) engages in spoliation of evidence, there is nothing inherently unfair about imposing sanctions against a litigant for the conduct of the litigant's agents.[62]

C. Contractual provisions

Although a party generally cannot disclaim liability for personal injury claims, defendants should be able to limit their potential liability for property damage claims.[63] As in most construction related cases, potential defendants can limit their liability by contract.[64] This includes (1) limitation of remedy provisions, (2) limitation of damages provisions, and (3) exclusion of incidental and consequential damages provisions.[65] There is no reason to believe that such disclaimers or limitations would not be upheld by courts in the mold context.

8.5 Damages

A. Attorneys' fees and costs

The general rule is that litigants bear their own expenses for fees and costs, except when specifically authorized by statute, rule, or agreement.[66] Attorney fee provisions in construction contracts should be carefully considered at the time of contracting.

When the New Jersey consumer fraud statute is implicated, defendants may be held liable for attorneys' fees and costs.[67] This is so, irrespective of express contract terms.

B. Cost of repair versus diminution of value

For property damage cases, the general rule is that the plaintiff is entitled to either (1) the cost to repair the damage or (2) the difference between the value of the property in its "damaged condition" versus the value of the property in its undamaged condition.[68] There is no reason to believe that the same rules will not apply in the mold context.

C. Opportunity to repair

Although many states now have "right to cure" laws (which permit builders to make repairs prior to the filing of a lawsuit), no such legislation currently exists in New Jersey.

Although there currently is no "right to cure" law in New Jersey mold-related litigation, there may be movement toward adopting such a law in the near future. Indeed, with respect to purchaser-builder arbitrations involving new construction, New Jersey's acting governor has issued an Executive Order toward that end. On May 9, 2005, New Jersey's Acting Governor Dick Codey signed Executive Order Number 33, which provides additional protections to new home buyers. The Order also creates a single arbitration hearing that will focus on defect repair instead of monetary relief. The standard, under the Order, is that the relief should be aimed at restoring the home to 'as new' condition. Thus, for new-home mold claims subject to arbitration, the builder may be provided an opportunity to repair as opposed to being subject to pure monetary damages.

D. Economic loss doctrine

The uncertainty created by the emerging nature of mold litigation can be partially diminished by the potential mold defendant's use of the "economic loss rule." The "economic loss rule" springs from the distinction between tort and contract.

Tort law compensates consumers for physical injury or damage to property other than the allegedly defective product itself. Contract law, on the other hand, compensates consumers for disappointed economic expectations. The economic loss rule maintains the distinction between tort and contract by requiring that claims for economic loss (i.e., disappointed economic expectations) be brought in contract. Thus, the "economic loss rule" is a judicially created doctrine that prohibits recovery in tort when a product has damaged only itself and the only losses suffered are economic in nature.[69]

This means that builders, building product manufacturers, or any other potential defendant in a mold case involving damage to the product itself (the structure at issue), should not be subject to the type of unlimited liability that is often associated with tort claims.

8.6 Admissibility of Scientific Evidence or Expert Testimony

The medical and scientific communities dispute over whether mold causes serious bodily injury. Accordingly, expert testimony on the relationship between mold and bodily injury is a significant element of any mold case. For expert testimony to be relevant, there must be a logical connection between the expert's theory and the expert's conclusion. With respect to mold litigation, this means that the expert must be able to demonstrate that exposure to mold caused the plaintiff's injury. Absent such proof, the expert's analysis is irrelevant.

To understand New Jersey's standard on the admissibility of expert testimony, one must be knowledgeable on two different standards that courts use for determining whether expert testimony is admissible—the *Frye* test and the *Daubert* test.[70] Pursuant to *Frye v. United States*, 293 F. 1013 (D.C. Cir. 1923), expert testimony is only admissible if it is based on scientific principles that are sufficiently established so that they have gained general acceptance in the relevant scientific community. Pursuant to *Daubert v. Merrell Dow Pharmaceuticals*, 509 U.S. 579 (1993), expert testimony is admissible only if the judge finds the testimony or evidence to be both reliable and relevant. Reliability is assessed by looking at (1) whether the theory or technique can and has been tested, (2) whether it has been subjected to peer review and publication, (3) the known or potential rate of error, and (4) whether the theory or technique has been generally accepted.[71]

New Jersey's courts use a hybrid of the *Frye* and *Daubert* standards. The admissibility of expert opinions is governed by New Jersey Rules of Evidence 702, 703, and 104. The New Jersey Supreme Court has charged trial courts with the responsibility in certain circumstances of conducting an evidentiary hearing pursuant to Rule 104 in cases in which "the scientific reliability of an expert's opinion is challenged."[72] The appropriate inquiry with respect to whether expert testimony is sufficiently reliable, is not whether the court thinks the expert's reliance on the underlying data was reasonable, but whether comparable experts in the field would actually rely on that information.[73] The Supreme Court held in *Rubanick v. Witco Chemical Corp.*, 125 N.J. 421, 593 A.2d 733 (1991), that a theory of causation that had not yet reached general acceptance in the scientific community "may be found to be sufficiently reliable if it is based on a sound, adequately founded scientific methodology involving data and information of the type reasonably relied on by experts in the scientific field."[74]

8.7 New Jersey Mold Legislation

On February 24, 2004, New Jersey Senate Bill 1249, entitled the "Toxic Mold Protection Act," was introduced in the Senate, and referred to the Senate Community and Urban Affairs Committee. If passed, this bill would require: (1) the Department of Health and Senior Services to adopt rules and regulations establishing permissible mold exposure limits and procedures for assessment of the health risks posed by the presence of mold; (2) the Department of Community Affairs to adopt mold identification guidelines and mold remediation procedures; and (3) the establishment of "mold disclosure" rules in the context of real estate transactions.[75]

Unlike mold disclosure laws of other states, New Jersey's mold disclosure statute does not provide the seller with immunity from litigation.[76]

8.8 Insurance Coverage

Mold claims and lawsuits may involve claims of negligent design and construction that are suitable for coverage under typical comprehensive general liability ("CGL") policies, builders' risk policies, and errors and omissions ("E&O") policies.[77] However, mold-related insurance claims are all too often met by the insurance industry's attempt to preclude coverage for these claims. With respect to policies issued prior to recent public awareness of mold related property and bodily injury claims, insurers often based their denial of coverage on the "absolute pollution exclusion." With respect to new policies being issued, insurers are inserting mold-specific exclusionary language into their policies.

A. Absolute pollution exclusion

Typical policy form exclusions for pollution bar insurance coverage for bodily injury or property damage arising out of the discharge, dispersal, release or escape of smoke, vapors, soot, fumes, acids, alkalis, toxic chemicals, liquids or gases, waste materials or other irritants or contaminants or pollutants into or upon land, the atmosphere or any water course or body of water.

It is pursuant to this language that insurers argue that the "absolute pollution" exclusion bars coverage for mold claims. New Jersey courts have held that the exclusion is not applicable to certain indoor pollutants.[78]

B. Mold specific exclusions

On September 11, 2003, the New Jersey Insurance Commissioner (Holly C. Bakke) issued a bulletin to all New Jersey P/C insurers giving specific guidelines on mold and fungus exclusions for personal and commercial lines. The bulletin, No. 03-24, replaces a previous set of guidelines contained in bulletin 02-14, issued July 8, 2002. The relevant portion of the announcement is as follows.

1. Personal lines and commercial lines property

Coverage should be offered at $10,000, on an aggregate basis, with optional increased limits of $25,000, and $50,000 made available. The $10,000 annual aggregate coverage should include loss to property caused by mold, fungi, wet or dry rot or bacteria, the cost to remove mold, fungi, wet or dry rot, or bacteria from property covered under Section 1, the cost to tear out and replace any part of the building or other covered property as needed to gain access to the mold, fungi, wet or dry rot, or bacteria; and, the cost of testing of air or property to confirm the absence, presence or level of mold, fungi, wet or dry rot, or bacteria. Coverage should apply only if such mold loss results from a "peril insured against" that occurs during the policy period. The limitation does not apply to losses due to fire and lightning.

2. Personal lines liability

Liability coverage should apply, up to at least $50,000, on an aggregate basis, for the total of all damages arising directly or indirectly, in whole or in part, out of the actual, alleged or threatened inhalation of, ingestion of, contact with, exposure to, existence of, or presence of any mold, fungi, wet or dry rot, or bacteria. Increased limits up to $100,000 should be available, when requested by the insured.

New Jersey's Department of Banking and Insurance's announcement also permits companies issuing "Commercial Lines General Liability" policies to provide coverage for mold up to the policy limits, provide sublimits, or exclude the coverage completely." It also states that insurers should continue "capturing Mold Data in the statistical format currently in use for other lines," and should notify policyholders of the options "to buy up to the higher limits."

8.9 Conclusion

This text highlights the nationwide inconsistency of proposed mold legislation, the unsettled spectrum of generally accepted mold remediation standards, and the unrestrained nature of personal injury verdicts. The result is the creation of a volatile landscape for those potential mold defendants who operate in more than one state. New Jersey does not currently possess a well-established body of law on mold litigation. Nonetheless, New Jersey has developed extensive bodies of case law on legal concepts readily applicable to mold cases. An understanding of the general principles presented in this chapter should not only serve as a primer for litigators confronted with mold claims, but should also provide guidance on how to counsel clients involved in such cases.

Endnotes

1. *See Gallara v. Koskovich, et al.*, 364 N.J. Super. 418, 431, 836 A.2d 840, 847 (Law Div. 2003) (stating that "Proof of negligence requires establishment of a duty of care and a breach thereof proximately causing the harm.").

2. *See* "Homeowner Says Water Problems Concealed in Disclosure Statement," *Mealey's Litig. Rep. Mold* (Vol. 3, Issue No.1, January 2003).

3. *See id.*

4. *See, e.g., Juliano v. Gaston*, 187 N.J. Super. 491 (App. Div. 1982), *cert. denied*, 93 N.J. 318 (1983) (relaxing the privity requirement in an action brought by homeowner against subcontractor).

5. *See Strawn v. Canuso*, 140 N.J. 43, 55-56, 657 A.2d 420, 426 (1995).

6. *See, e.g., Dupler v. Schwager*, 2003 WL 22429101 (Oh. App. 2d Dist. Oct. 24, 2003) (caveat emptor does not apply to latent defects); *Ray v. Clark*, 2000 WL 1873902 (Wash. App. Div. 2000) (finding an exception to caveat emptor when the landlord fails to disclose the existence of a latent defect).

7. *Id.*

8. *Id.*

9. *See* 2004 N.J. Senate Bill 1249 (211 Legislature).

10. *See Aronsohn v. Mandara*, 98 N.J. 92, 105-106, 484 A.2d 675, 682–683 (1984).

11. *See* N.J.S.A. 46:3B-2; *see also, Ingraham v. Trowbridge Builders*, 297 N.J. Super. 72 (App. Div. 1997).

12. *Id.*

13. *Id.*

14. *See* N.J.S.A. 46:3B-9.

15. *See* Melvon Aron Eisenberg, "Third-Party Beneficiaries," *Columbia L. Rev.* 92:1358, 1402–1403 (1992).

16. *See* Steven B. Lesser and Amy L. Koltnow, "Liability of Construction Design Professionals and the Economic Loss Rule," *WTR Brief* 29:50 (2000).

17. *Juliano v. Gaston*, 187 N.J. Super. 491 (App. Div. 1982), *cert. denied*, 93 N.J. 318 (1983).

18. *See id.*; It was noted that in *Totten v. Gruzen*, 52 N.J. 202 (1968) the New Jersey Supreme Court repudiated the "completed and accepted" rule, which was a historical bar to an action by the victim of defective workmanship after the owner's acceptance of the completed project.

19. *Id.* at 494.

20. *Id.*

21. It should be noted that the plaintiffs first instituted the same cause of action against the prime contractors and received a default judgment that was never satisfied because of the prime contractor's insolvency. Thus, while some may argue this is a case of "corrective justice," and should be limited as such, there is no such restrictive language in the opinion.

22. *Id.* at 495, 497.

23. *Id.* at 497.

24. *Id.* at 497-498.

25. *Id.*

26. *See Coyle v. Englander's*, 199 N.J. Super. 212, 488 A.2d 1083 (App. Div. 1985).

27. *See McDonald v. Mianecki*, 159 N.J. Super. 1, 386 A.2d 1325 (App. Div. 1985); *Ramapo Brae Condominium Ass'n, Inc. v. Bergen County Housing Authority*, 328 N.J. Super. 561, 746 A.2d 519 (App. Div. 2000); *Aronsohn v. Mandara*, 98 N.J. 92, 484 A.2d 675 (1984).

28. *See Aronsohn v. Mandara*, 98 N.J. 92, 99, 484 A.2d 675, 678-679 (1984).

29. *Id.*

30. *Id.*

31. *Id.*

32. *See Reste Realty Corp. v. Cooper*, 53 N.J. 444, 251 A.2d 268 (1969).

33. *Id.*

34. See *McCurdy v. Wyckoff*, 73 N.J.L. 368, 63 A. 992 (Sup. Ct. 1906).

35. *See, e.g., Higgins v. Whiting*, 102 N.J.L. 279, 131 A. 879 (Sup. Ct. 1926).

36. *See Winslow v. Corporate Express, Inc.*, 364 N.J. Super. 128, 140, 834 A.2d 1037, 1044 (App. Div. 2003); *Zorba Contractors, Inc. v. Housing Auth. of the City of Newark*, 362 N.J. Super. 124, 139, 827 A.2d 313, 322 (App. Div. 2003); *Richie and Pat Bonvie Stables, Inc. v. Irving*, 350 N.J. Super. 579, 589, 796 A.2d 899, 906 (App. Div. 2002).

37. *See Kuhnel v. CNA Insurance Companies*, 322 N.J. Super. 568, 581, 731 A.2d 564, 571 (App. Div. 1999); *Carroll v. Cellco Partnership*, 313 N.J. Super. 488, 502, 713 A.2d 509, 516 (App. Div. 1998).

38. *See, e.g., Strawn v. Canuso*, 140 N.J. 43, 55–56, 657 A.2d 420, 426 (1995) (holding that under certain circumstances a professional seller of real estate's "silence may be fraudulent").

39. *See* N.J.S.A. 56:8-1 et seq.; *Strawn v. Canuso*, 140 N.J. 43, 55–56, 657 A.2d 420, 426 (1995).

40. *See Strawn v. Canuso*, 140 N.J. 43, 60-61, 657 A.2d 420, 428–429 (1995).

41. *Id.*

42. *Id.*

43. *See Strawn v. Canuso*, 140 N.J. 43, 60-61, 657 A.2d 420, 428–429 (1995).

44. Diego Graglia, "Teacher Gets $61,000 in Mold-Related Case," the *Star Ledger* (February 13, 2004).

45. *Id.*

46. *See Strawn v. Canuso*, 140 N.J. 43, 55-56, 657 A.2d 420, 426 (1995).

47. *See Greczyn v. Colgate Palmolive, et al.*, 367 N.J. Super. 385, 390-391, 842 A.2d 895, 898 (App. Div. 2004).

48. *Id.*

49. *See Cyktor v. Aspen Manor Condominium Ass'n et. al.*, 359 N.J. Super. 459, 470, 820 A.2d 129, 137 (App. Div. 2003).

50. N.J.S.A. 2A:14-1.1.

51. *Id.*

52. *Wayne Township Board of Education v. Strand Century, Inc., et al.*, 172 N.J. Super. 296 (App. Div. 1980); *see also, Hall v. Luby Corp.*, 232 N.J. Super. 337 (Law Div. 1989).

53. *Rolnick v. Gilson & Sons, Inc.*, 260 N.J. Super. 564, 567-68, 617 A.2d 288 (App. Div.1992).

54. *See Wayne Township Board of Education v. Strand Century, Inc., et al.*, 172 N.J. Super. 296 (App. Div. 1980); *see also, Hall v. Luby Corp.*, 232 N.J. Super. 337 (Law Div. 1989).

55. *See, e.g., Manorcare Health Services, Inc. v. Osmose Wood Preserving, Inc.*, 336 N.J. Super. 218, 226 (App. Div. 2001).

56. *Manorcare Health Services, Inc. v. Osmose Wood Preserving, Inc.*, 336 N.J. Super. 218, 226 (App. Div. 2001).

57. *Id.*

58. *Id.* at 226; Aetna Life and Casualty Co., 309 N.J. Super. 358, 365 (App. Div. 1998).

59. *Hirsch v. General Motors*, 266 N.J. Super. 222, 251, 264 (Law Div. 1995) (holding that "a report prepared at plaintiff's request with prospective litigation in mind without affording defendant's experts an opportunity to inspect the same best evidence and prepare a defense is substantially prejudicial to defendants").

60. Manorcare, 336 N.J. Super at 226; *see also,* Aetna, 309 N.J. Super. at 368; Hirsch, 266 N.J. Super. at 256.

61. See *id.*

62. *Hewitt v. Allen Canning Company*, 321 N.J. Super. 178, 185 (App. Div. 1999) (*citing Nerney v. Garden State Hospita*l, 229 N.J. Super. 37 (App. Div. 1988)).

63. *See, e.g., Alloway, III v. General Marine Indust., L.P.*, 149 N.J. 620, 630, 695 A.2d 264, 269 (1997) (stating that "the U.C.C. allows manufacturers to limit their liability through disclaimers, except for personal injuries"); *Saltiel v. GSI Consultants, Inc.*, 170 N.J. 297, 788 A.2d 268 (2002) (stating that Prosser & Keeton distinguish between tort and contract claims through the rule that "tort obligations may not be disclaimable"); *New Mea Construction Corp. v. Harper*, 203 N.J. Super. 486, 494, 497 A.2d 534, 538 (App. Div. 1985) (also stating that Prosser & Keeton distinguish between tort and contract claims through the rule that "tort obligations may not be disclaimable").

64. *See, e.g., Marboro, Inc. v. Borough of Tinton Falls*, 297 N.J. Super. 411, 417, 688 A.2d 159, 162 ("Our courts have traditionally upheld contractual limitations of liability").

65. *See, e.g., General Motors Acceptance Corp. v. Jankowitz*, 216 N.J. Super. 313, 327, 523 A.2d 695, 702 (App. Div. 1987) (a seller may limit or exclude consequential damages); N.J.S.A. 12A:2-719 Contractual modification or limitation of remedy.

66. *Velli v. Rutgers Casualty Ins. Co.*, 257 N.J. Super. 308, 309 (App. Div.), *cert. denied*, 130 N.J. 597, 617 A.2d 1220 (1992); J*osantos Const. v. Bohrer*, 326 N.J. Super. 42 (App. Div 1999).

67. *See* N.J.S.A. 56:8-1 et seq.

68. *See Berg v. Reaction Motors Division, Thiokol Chemical Corp.*, 37 N.J. 396, 412, 181 A.2d 487, 495 (1962) ("In its treatment of the measure of damages when harm is done to personal property, the Restatement of Torts provides that the plaintiff be given an option to recover either the diminution in value or the reasonable cost of repairs.").

69. *See Goldson v. Carver Boat Corp.*, 309 N.J. Super. 384, 707 A.2d 193 (App. Div. 1998).

70. *Frye v. United States*, 293 F. 1013 (D.C. Cir. 1923); *Daubert v. Merrell Dow Pharmaceuticals*, 509 U.S. 579 (1993).

71. *Daubert v. Merrell Dow Pharmaceuticals*, 509 U.S. 579 (1993).

72. *Kemp ex rel. Wright v. State*, 174 N.J. 412, 425–27, 809 A.2d 77 (2002).

73. N.J.R.E. 702.

74. *See Rubanick v. Witco Chemical Corp.*, 125 N.J. 421, 449, 593 A.2d 733, 747-48 (1991).

75. *See* 2004 N.J. Senate Bill 1249 (211 Legislature).

76. *See, e.g.*, The "Montana Mold Disclosure Act" Montana Code. Ann., Tit. 70, Ch. 16, Pt. 7 § 701 et seq. (2003).

77. It should be noted that in general ten states have statutes that provide new homeowners with what is sometimes referred to as homeowner warranty insurance or guarantees which are provided through a new home builder or a state sponsored program. *See generally* Connecticut (C.G.S.A. § 47-116); Florida (F.S.A. § 634.327); Indiana (IC 32-15-7-1); Louisiana (LSA-R.S. 9:3141); Maryland (Code, Real Property, § 10-601); Mississippi (Miss. Code Ann. § 83-58-1); New Jersey (N.J.S.A. 46:3B-1); South Carolina (Code 1976 § 40-29-190); Virginia (Code 1950, § 55-70.1). In general, homeowner warranty programs may impose certain time restrictions for claims and limit a builder's liability to the purchase price of the home should a new homeowner elect to participate in the program. *See, e.g.,* N.J.S.A. 46:3B-4, 46:3B-9.

78. Byrd *ex rel. Byrd v. Blumenreich*, 317 N.J. Super. 496, 722 A.2d 598 (App. Div. 1999) (holding that "Penn National's absolute pollution exclusion clause is uncertain or ambiguous and susceptible to differing interpretations as to whether a claim for personal injury caused by the *indoor* residential ingestion of lead paint chips, flakes, or dust is excluded from coverage. Such uncertainty or ambiguity is fatal to Penn National's position.") (emphasis added). *NAV-ITS, Inc. v. Selective Ins. Co. of America*, 183 N.J. 110, 121 (2005) (finding Pollution Exclusion to be limited to "traditional environmentally related damages," which did not include fumes from floor sealant).

Chapter 9

Mold in Pennsylvania

Douglas F. Schleicher, Esq., and Bryan D. Weiss, Esq.

Synopsis

9.1 Introduction

There have not been a significant number of enforcement actions or lawsuits in Pennsylvania relating to mold. This chapter will provide a general overview of the mechanisms that may be available to government agencies at the state and local level to regulate mold related exposure. In addition, various legal theories of liability in private actions, based on both statutory and common law, will be discussed in order to help the reader anticipate where private litigation may be focused in the future.

9.2 State and Local Enforcement Authority

To date, Pennsylvania has not enacted legislation specifically addressing mold. Notwithstanding the foregoing, there are several different governmental bodies which may have enforcement authority relating to the presence of mold.

A. Pennsylvania Department of Environmental Protection (PADEP)

Pursuant to several different sections of the Administrative Code of 1929, PADEP has broad authority to "protect the people of [Pennsylvania] from unsanitary conditions and other nuisances." 71 P.S. § 510-17. As part of this broad power grant, the legislature has provided PADEP with the power to examine and survey property in the state with respect to "nuisances." *Id.* If such conditions are identified, PADEP can then order the alleged nuisance to be abated or removed. *Id.* Thereafter, if the owner or occupant of the property fails to respond to the order, PADEP also has the statutory right to enter the premises, abate the nuisance itself and later seek recovery of its costs in doing so. *Id.* With respect to tenements, lodging or boarding houses, PADEP also has the authority to condemn the property for unsanitary conditions. 71 P.S. § 510-19.

B. Local cities and municipalities

The Pennsylvania legislature has enacted several laws authorizing cities and municipalities to promulgate local regulations regarding the sanitation and inspection of all forms of improved real estate. With this grant of power most local entities have enacted local housing and building regulations requiring that properties be kept in a sanitary condition. Most of these local regulations empower local sanitary officers with entry, inspection and abatement authority similar to those available to PADEP. Local regulations also generally provide that a local municipality has the right to condemn properties that constitute an ongoing public nuisance or are otherwise uninhabitable.

9.3 Potential Private Rights of Action

To date, there has been minimal mold litigation in Pennsylvania. However, the following will summarize some of the relevant private rights of action that may be applicable.

A. Contractual claims for economic loss

The specific terms of a contract will govern any contract claim. Absent a provision pertaining to mold or sanitation, certain types of claims can be expected. Economic loss may include failure to receive fair value (e.g., unsanitary housing or commercial space) in real property transactions and lost profits from down time associated with the presence of mold, mold enforcement actions or remediation programs.

1. Implied warranty of habitability (residential)

In the absence of express contractual provisions, residential leases are subject to an implied warranty of habitability. *Pugh v. Holmes*, 405 A.2d 897 (Pa. 1979). Therefore, the landowner has an implied obligation to provide a tenant with improvements that are safe and sanitary, even if the lease language itself does not explicitly speak to such issues.

Although a local housing code violation does not, by itself, establish a breach of the warranty, such violations can generally be used as evidence to show a material breach of the obligation. *See Pugh, supra* at 906. The materiality of a landlord's breach is a question of fact and the factors that Pennsylvania courts have traditionally proffered for assessment of whether or not a material breach has occurred include the following: does the condition violate a housing code, law or regulation; the nature and seriousness of the condition; the effect of the defect on safety and sanitation; the length of time during which the defect has persisted; and the age of the improvement.

In addition to proving that a material breach has occurred, the plaintiff will also be required to show that it gave notice to the landlord of the defect or condition; the landlord had a reasonable opportunity to make the necessary repairs; and that the landlord failed to do so. *Id.*

Assuming that a plaintiff is successful in proving breach of the habitability obligations, standard contractual remedies apply. Recoverable economic damages include rent rebates and amounts spent on reasonable repairs to make the property "habitable" again. In addition, at least one federal court sitting in Pennsylvania has authorized a damage award for "deprivation and humiliation" related to a breach of warranty of habitability claim. *See In re Clark*, 906 B.R. 569 (Bankr. E.D. Pa. 1989).

2. Implied warranty of habitability (commercial)

As a general matter, Pennsylvania courts have been unwilling to read a warranty of habitability into commercial leases because the leases are assumed to be the result of arms length negotiations between parties with roughly equal bargaining power. *Hirsh v. Carbon Lehigh Intermediate Unit #21*, 65 Pa. D.&C. 4th (Pa. Com. Pl. 2003).

Pennsylvania courts have, however, also held that general contract law no longer binds commercial lessees to the "independence of obligations" doctrine, under which a tenant had only one remedy if the landlord was in material breach of the lease: surrender the premises and be relieved of the obligation to pay rent. Under more recent Pennsylvania case law, at least two other offset-type options exist, including performing remediation on behalf of the landlord and reducing rent for the costs associated with same; or deducting from the rent the difference between the rental value of the premises as leased and the value of the property if the lease had been complied with (i.e., rent abatement). *See Pawco, Inc. v. Bergman Knitting Mills, Inc.*, 424 A.2d 891 (Pa. Super. 1980).

In light of the foregoing, sophisticated commercial tenants should seek affirmative representations, warranties and covenants from the landlord regarding the sanitary condition of their space and other common areas of the building. In addition, they should consider identifying in the lease specific remedies that would be available in the case of a breach.

3. Implied warranty of construction in a good and workmanlike manner

Under Pennsylvania law, construction contractors are deemed to provide an implied warranty that construction was performed in a reasonably workmanlike manner and that the finished product is fit for the purpose intended. *Elderkin v. Gaster*, 288 A.2d 771, 777 (Pa. 1972). This warranty is based in contract law and provides regular contract damages available to the plaintiffs. However, there is no contractual privity requirement in order to assert such a claim. *Spivak, supra* at 405. Pennsylvania courts have also held that the warranty of workmanship will be implied with respect to both residential and commercial properties. *Pittsburgh Nat'l Bank v. Welton Becket Associate et al.*, 601 F. Supp. 887 (W. D. Pa 1985); *Sports Management Group, Inc., supra* at 765.

B. Breach of lease

At least one court in Pennsylvania has interpreted specific lease provisions as they pertain to the presence of mold in a building. In *Hirsh v. Carbon Lehigh Intermediate Unit #21*, 65 Pa. D.&C. 4th (Pa. Com. Pl. 2003), a school district terminated a lease with its landlord due to a recurring mold condition. In response, the landlord filed suit asserting that the school district had breached its non-residential lease agreement and sought to recover unpaid rent for the remainder of the lease term. The court held that the presence of mold did not constitute a breach of the casualty provision of the lease because it did not totally destroy or extensively damage the premises, notwithstanding that the school was vacated during the school year for a short period of time to allow for remediation. The court found that the school district bore part of the responsibility for the mold's presence, since it decided not to install an HVAC system as part of the construction of the building. Further, the court noted that the school district failed to notify the landlord of the mold's recurrence, which precipitated the district's unilateral attempt to terminate the lease. The court held that the district breached its lease, and awarded damages to the landlord.

C. Pennsylvania Unfair Trade Practices and Consumer Protection Law

Pennsylvania courts have held that the Unfair Trade Practices and Consumer Protection Law ("UTPCPL") provides a private right of action related to sales and leases of real estate. *See Skurnowicz et al. v. Lucci et al.*, 798 A.2d 788 (Pa. Super. 2002); *Commonwealth of Pennsylvania v. Monumental Properties, Inc., et al.*, 329 A.2d 812 (Pa. 1974). The act prohibits "unfair or deceptive acts or practices", which are further defined by a specific list of activities, including actual and implied misrepresentations to the public. 73 Pa.C.S. §201-2(4). The elements of such a case are the same as those for a common law fraudulent misrepresentation claim: there must be a material representation; made with knowledge of its falsity, or recklessness as to its truthfulness; an intent to mislead the aggrieved party; justifiable reliance on the misrepresentation; and resulting injury that was proximately caused by the justifiable reliance. *Skurnowicz, supra* at 793. If a property owner is aware of mold growth on the property and fails to disclose it to the purchaser or lessee, a seller may be able to craft a credible argument that there has been a violation of the act.

The statute generally provides that a private plaintiff can seek redress for "any ascertainable loss of money or property" that results from a violation of the act. 73 Pa C.S. §201-9.2(a). In addition to traditional tort remedies, the UTPCPL also provides trial courts with authority to grant up to three times the actual damages sustained, along with associated costs and reasonable attorney fees.

Finally, the UTPCPL has been extended to reach more than fraud. A federal bankruptcy court sitting in Pennsylvania has held that evidence of a continuous breach of the implied warranty of habitability by a landlord can be the basis of a violation of UTPCPL. *See In re Clark*, 96 B.R. 569 (Bankr. E.D. Pa. 1989). Under *Clark*, fraud need not be proven to make out a successful claim under the UTPCPL. While no Pennsylvania state court has directly addressed this issue, and the case involved a variety of damning facts, including that the landlord did not follow court orders to improve the condition of the living spaces, *Clark* signals an expansion of the situations in which the UTPCPL applies.

D. The Real Estate Seller Disclosure Law

The Real Estate Seller Disclosure Law (RESDL), 68 Pa.C.S.A. § 7301 *et seq.*, was enacted in December, 2000. It pertains to sales of residential properties of one to four units. 68 Pa.C.S.A. § 7302(a); 68 Pa.C.S.A. § 7103(a). The RESDL requires sellers of subject properties to disclose to the buyer all known material defects pursuant to a form promulgated by the State Real Estate Commission. 68 Pa.C.S.A. §§ 7303-7304(a). Although there is no required disclosure on the form relating to mold, there are required disclosures relating to water, roof and related building conditions and hazardous substances, many of which may give rise to a required disclosure pertaining to mold. A real estate transfer covered by RESDL may not be invalidated solely because of a failure by a party to comply, but that party is subject to liability for the actual damages suffered by the buyer as a result of any violation, and a court may impose such additional damages or remedies as may otherwise apply. 68 Pa.C.S.A. § 7311(a).

E. Torts

Pennsylvania courts have yet to render any significant decisions in cases addressing the alleged health effects related to mold exposure. However, personal injury and property damage suits against property owners or property management companies in Pennsylvania could be premised on several well-recognized tort theories. In addition to property owners, entities such as design professionals (i.e. architects and engineers) and contractors may be targeted in suits based on allegedly defective construction or design of buildings or HVAC systems.

1. Negligence (premises liability)

Mold personal injury and property damage cases can arise under many different factual scenarios. One frequent context involves premises liability. Pennsylvania recognizes premises liability claims based in negligence. A defendant's standard of care usually depends on its relationship to the plaintiff. For example, differing standards of care are applied to licensees, public invitees and business invitees of the entity who owns or controls the property. *See Cresswell v. End*, 831 A.2d 673 (Pa. Super2003); *Gutteridge v. A.P. Green Services, Inc. et al.*, 804 A.2d 643 (Pa. Super. 2002).

Although a review of Pennsylvania case law does not indicate any reported cases involving tenant personal injury claims based exclusively on mold exposure, there are pertinent cases involving indoor air quality. In *Wright, et al. v. McDonald's Corp.*, No. 91 Civ. 2061, 1993 WL 53582 (E.D.Pa. March 2,1993), the plaintiffs alleged various personal injuries as a result of the inhalation of carbon monoxide at the franchised premises that they leased from McDonald's. The court noted that under Pennsylvania law a landlord out of possession is generally under no duty to maintain and repair leased premises. However, the landlord has such a dut when (1) the dangerous condition existed at the time of the leasing; (2) t nd-lord knew or should have known of the condition; and (3) the tenan not have discovered the condition with a reasonable examinati *Wright*, the judge denied the defendant's motion for summary judg finding that the plaintiffs had shown sufficient evidence that McDo designed and constructed a hot water heater ventilation system wi tent defect of which it should have been aware and that the plaintif not have discovered the defect upon reasonable inspection.

Although *Wright* does not deal specifically with mold, it is instructive because mold exposure also can result from inhalation of spores generated by defective building systems or conditions not susceptible to discovery.

A residential or commercial landlord may be subject to liability if it has knowledge of a mold condition in a space to which the tenants do not have access, such as the basement, utility or other areas. Depending on the design of the particular building or the equipment in it, air may also be drawn from the basement or ground floor areas and dispersed throughout the building thereby potentially providing an exposure pathway to tenants. Evidence of such circumstances could establish a duty upon a landlord or property manager to respond to mold problems that arise as a result.

In *William Penn School District v. Workers' Compensation Appeal Board*, 717 A.2d 589 (Pa. Commw. 1998), the Commonwealth Court upheld the decision of the workers' compensation judge awarding damages to an instructional assistant who claimed disabilities resulting from exposure to bacteria and fungus in the school basement classroom in which she worked. The workers' compensation judge found that the claimant had sustained work related injuries in the nature of an aggravated and exacerbated previous asthmatic condition and that she was totally disabled. While the claimant's own expert testified that there were a number of contributing causes, including a history of smoking, the Commonwealth Court determined that there was "substantial evidence" in the record to support the workers' compensation judge's findings. William Penn School District was ordered to pay wage loss benefits and outstanding medical bills.

2. Professional negligence

Personal injury claims attributable to mold may arise as a result of the negligence of architects or engineers in the design or supervision of construction of a building, its HVAC systems, building envelope or other features. *See Bonnette, et al. v. Tunica-Biloxi Indians*, No. 02-919, 2003 WL 9282 (La. Ct. App. May, 28, 2003) (several groups of plaintiffs filed against construction defendants, including an architect and an engineering firm, for personal injuries from alleged mold exposure at a casino related child care facility). The Superior Court of Pennsylvania has ly recognized that "an architect may be held liable for negligence in failing to exercise the ordinary skill of his profession, which results in the erection of an unsafe structure whereby anyone lawfully on the premises is injured." *Young v. Eastern Engineering and Elevator Company, Inc.*, 554 A.2d 77, 79 (Pa. Super. 1989); 97 A.L.R.3d §2(a) (1980); 5 Am.Jur 2d, Architects §25; *see also Prost v. Caldwell Store, Inc.*, 187 A.2d 273, 276 (Pa. 1963). Other Pennsylvania cases confirm that design professionals have a duty to unrelated third parties for damages arising from faulty building designs. *See, e.g., McKonley v. Franco*, 12 Pa. D. & C.3d 105 (1979) (subsequent owners of building were able to maintain suit against architect in negligence after roof collapsed without privity of contract). Theoretically, this line of cases support a claim asserting negligent design that results in injury or damage due to mold.

The duty of the design professional was addressed in the context of a property damage claim in *City of Allentown v. O'Brien and Gere Engineers, Inc.*, 1997 WL 256060 (E.D.Pa.). In *City of Allentown*, the federal district court was charged with deciphering "the conduct expected of a reasonably competent engineering firm with respect to recommending, designing and supervising the construction of a structure appropriate for the client's needs." The court provided three qualitative measures for fact finders to use when assessing the conduct of a design professional: (1) did the professional use the reasonable care and diligence expected of a design professional in considering and evaluating all information pertinent to the project undertaken; (2) did the professional apply the knowledge reasonably expected of a design professional to make appropriate design recommendations to the client in accordance with sound design principles; and (3) did the professional apply the skill and knowledge reasonably expected of a design professional to design a structure or facility that is reasonably likely to accomplish the purpose for which it was retained. *Id.; Bloomsburg Mills v. Sordoni Construction Co.*, 164 A.2d 201 (Pa. 1960).

3. Intentional infliction of emotional distress

Mold exposure may provide grounds for intentional infliction of emotional distress ("IIED") claims. In the residential housing context, Pennsylvania courts have refused to dismiss the theory that evidence of uninhabitable conditions at a leased premise can be used as the basis of a claim for IIED. *See Fair v. Negley*, 390 A.2d 240 (Pa. Super. 1978); *Beasley v. Freedman*, 389 A.2d 1087 (Pa. Super. 1978). The elements of a claim of

IIED include proving that the landlord must have acted intentionally or with reckless disregard of the consequences of its acts; the landlord's conduct must have been extreme or outrageous; the lessee must have suffered severe emotional distress; and the landlord's conduct was the cause of the emotional distress. In *Fair*, the court noted that the plaintiffs had adequately pled a case for IIED and that they would have an opportunity to prove that the defendant's actions in allegedly breaching the warranty of habitability was the kind of outrageous behavior that could make out a claim for IIED. Although the *Fair* case did not involve mold, it arose from indoor air quality problems allegedly caused by an improperly ventilated gas hot water system and space heaters on the premises.

4. Strict products liability

In addition to targeting the entities that own, manage, design or construct buildings, personal injury plaintiffs have also targeted manufacturers of exterior cladding systems using strict product liability theories. *See Gropper, et al. v. STO Corporation, et al.*, 552 S.E.2d 118 (Ga. App. 2001) (homeowners filed suit against builder, stucco subcontractor and stucco manufacturer asserting claims of strict products liability for mold-related personal injuries and property damage due to use of synthetic stucco product on the exterior of their home).

Pennsylvania courts have historically recognized an action of strict products liability under Section 402A of the Restatement (Second) of Torts (1965). *See Webb v. Zern*, 200 A.2d 853 (Pa. 1966). The elements of a claim for strict products liability are that the product was sold in a defective condition unreasonably dangerous to the public; that the defect existed when it left the hands of the defendant; and that the defect caused harm to the plaintiff. *See DeSantties v. Frick Co.*, 745 A.2d 624 (Pa. Super. 2000). As a practical matter, the "unreasonably dangerous" requirement was deemphasized by the Pennsylvania Supreme Court in *Berkebile v. Brantly Helicopter Corp.*, 337 A.2d 893 (Pa. 1973), wherein the court noted that this qualification was only provided to prevent a manufacturer from being held liable where the product is not defective. Accordingly, once a product is shown to be defective and that the defect has caused injury, the "unreasonably dangerous" requirement has been met. *Id.*

Pennsylvania courts have regularly followed § 402A and applied the doctrine to property damages. *See Cumberland Valley Joint School Author-*

ity v. Haldeman, Inc., 23 Pa. D. & C.3d 616 (1982); *2-J Corporation v. Tice Construction Co. et al.*, 126 F.3d 539 (3rd Cir. 1997). In *Cumberland Valley*, a roof coating of "silicone asbestolite" was defective and caused water ingress and resultant damage to the plaintiff's roof. The defendant wholesaler asserted that the claim was barred by the economic loss doctrine. The court denied the defendant's motion to dismiss noting that the plaintiff had properly alleged resultant property damage. *Id.* Although it is not a mold case, *Cumberland Valley* provides guidance and support for future mold plaintiffs who can trace their problems to water ingress or mold growth directly attributable to a defective building product.

5. Medical monitoring damages

In addition to personal injury damages such as medical expenses and lost earnings, Pennsylvania has recognized a separate claim for medical monitoring costs. *See Redland Soccer Club, Inc. v. Department of the Army, et al.*, 696 A.2d 137(Pa. 1997). In *Redland*, the court noted that a common law claim for medical monitoring exists if a toxic tort plaintiff proves: 1) an exposure greater than normal background levels; 2) to a proven hazardous substance; 3) caused by the defendant's negligence; 4) as a proximate result of the exposure the plaintiff has a significantly increased risk of contracting a serious latent disease; 5) a monitoring procedure exists that makes the early detection of the disease possible; 6) the prescribed monitoring regime is different from that normally recommended in the absence of the exposure; and 7) the prescribed monitoring regime is reasonably necessary according to contemporary scientific principles.

In *Heller v. Shaw Industries, Inc.*, 1997 WL 535163 (E.D.Pa.), *aff'd* at 167 F.3d 146 (3rd Cir. 1999), plaintiffs alleged personal injury (sick building syndrome) caused by excessive levels of volatile organic compounds ("VOC") that were released into the air in the plaintiffs' home from carpeting manufactured by the defendant. The plaintiffs asserted a claim for medical monitoring damages. However, the trial judge made the following determination:

> … plaintiffs have adduced no evidence that defendant's carpets emitted VOCs at concentrations harmful to health, that the levels of VOCs detected by [the consultant] are higher than the normal background presence, that plaintiffs suffer increased risk of contracting a serious latent disease as a result of their exposure to the levels of VOCs detected by

> [the consultant], or that the medical monitoring would be beneficial to the treatment and early detection for a serious latent disease. *Id.* at 21.

Although the *Heller* case does not specifically address mold, it is instructive as to the difficulty a mold plaintiff may have in trying to make out a claim for medical monitoring expenses. At this time, there is little conclusive, peer-reviewed evidence that mold is generally toxic, that it is the cause of any serious latent disease or that an expert can predict that individuals will be negatively affected by it.

6. Negligent misrepresentation (professional)

Mold litigation often involves finger pointing between contractors and professionals with respect to responsibility for the specific conditions leading to mold growth. Contractors may allege defects in the project's design. This can be done via claims based on the Restatement (Second) of Torts §552 arguing that the design professional's plans and specifications incorporated "negligent misrepresentations." In order to state a claim for negligent misrepresentation, a contractor must show misrepresentation of a material fact; that the defendant either knows it is a misrepresentation, makes it without knowledge of its truth or falsity, or makes the representation under circumstances in which it ought to know of its falsity; that the defendant must intend the representation to induce another to act on the misrepresentation; and that the contractor suffered an injury acting in justifiable reliance on the misrepresentation. *See Borough of Landsdowne v. Sevenson Environmental Services, Inc. et al.*, 2000 WL 1886578 (E.D.Pa.) *citing Williams Controls, Inc. v. Parente, Randolph et al.*, 39 F. Supp.2d 517 (M.D.Pa. 1999).

At this time, there are no reported cases in Pennsylvania involving a contractor's claims of negligent misrepresentation against a design professional related to mold. It is also unclear whether a contractor will ultimately be successful making these claims because of the economic loss doctrine, which bars tort claims seeking exclusively economic damages. For example, a property owner may choose to sue only the contractor for its mold-related property damages. The contractor may, thereafter, attempt to add the design professional as a third-party defendant, alleging negligent misrepresentation. Such a claim would likely be barred by the economic loss doctrine. *David Pflumm Paving & Excavating, Inc. v.*

Foundation Services Company, 816 A.2d 1164, 1171 (Pa. Super. 2003); *but also see Borough of Lansdowne, supra* (economic loss doctrine not applicable to negligent misrepresentation claims).

7. Implied warranty of fitness for intended purpose (professionals)

Although a contractor's action for negligent misrepresentation against an architect may be precluded by the economic loss doctrine, design professionals may find it difficult to extricate themselves from the net of liability. Typically, a design professional impliedly warrants that its plans and specifications will result in a structure reasonably fit for its intended purpose. *See, e.g., Bloomsburg, supra at 203; Metropolitan Edison Co v. United Engineers & Constructors, Inc.*, 4 Pa. D & C.3d (1977). It has been determined that third parties may rely on this implied warranty. *See, e.g., Spivak v. Berks Ridge Corporation Inc. et al.*, 586 A.2d 402, 405 (Pa.Super.1991). Therefore, a contractor can be expected to assert a breach of implied warranty claim against a professional arguing that the mold-related damage is the result of the professional's errors or omissions.

8. Negligent construction

Contractors and construction managers are likely to be targeted directly by owners in property damage claims related to mold and moisture conditions. *See Centex-Rooney, supra* (Florida appellate court upheld $8.8 million dollar breach of contract award against construction manager for failing to properly supervise construction). To date, there are no reported Pennsylvania opinions discussing claims that can be brought against contractors arising from mold. However, Pennsylvania courts generally recognize several different theories of redress against contractors for construction defects, including an action for "negligent construction." *See Spivak, supra* at 405 (contractor has a common law duty to perform its work in a reasonable manner); *see also Sports Management Group, Inc. v. Allensville Planning Mill, Inc.*, 16 Pa. D. & C.3d 760 (1980).

9. Intentional misrepresentation/failure to disclose (sale of property)

Intentional misrepresentation is actionable if the following is shown: (1) a representation; (2) which is material to the transaction at hand; (3) made falsely, with knowledge of its falsity or recklessness as to whether it is true or false; (4) with the intent of misleading another into relying on it; (5) justifiable reliance on the misrepresentation; and (6) the resulting injury was proximately caused by the reliance. *Bortz v. Noon*, 729 A.2d 555, 560 (Pa. 1999). The tort of intentional non-disclosure has the same elements as intentional misrepresentation, except that with non-disclosure the party intentionally conceals, rather than affirmatively misrepresents, the material fact. *Id.* While concealment may be fraudulent, mere silence is not sufficient in the absence of a duty to speak. *Sevin v. Kelshaw*, 611 A.2d 1232, 1236 (Pa. Super. 1992); *Smith v. Renaut*, 564 A.2d 188, 192 (Pa. Super. 1989). A real estate broker, in addition to a seller, can be liable for the torts of fraud and intentional non-disclosure. *Bortz v. Noon*, 729 A.2d at 560.

Historically, fraudulent concealment has not been actionable unless there is active concealment. *See Anderson v. Harper*, 622 A.2d 319, 323 (Pa. Super. 1993). However, in *Roberts v. Barbagallo*, 531 A.2d 1125 (Pa. Super. 1987), a far-reaching decision with potentially significant ramifications, the Superior Court held that liability for fraudulent concealment extends to failure to provide generally available information about an allegedly toxic substance. In Roberts, the court held that a broker was liable for fraudulent concealment because it failed to disclose to the buyer material information concerning the possible health effects from urea formaldehyde foam insulation ("UFFI") and thereby prevented the buyer from investigating the insulation in the house prior to her signing the agreement of sale. *Id.* at 1131. The court found that the information concerning UFFI was material because there was a federal ban on UFFI, there were known potential health risks from the use of UFFI, and the local board of realtors had recommended that information concerning UFFI was material to real estate transactions, should be disclosed to both buyers and sellers and that an effort should be made to try to ascertain the use of UFFI in properties being listed for sale, all of which was known to the broker. *Id.* The court does not satisfactorily explain how the failure to disclose the material information prevented the buyer from investigating UFFI in the house, although the court's decision may be attributable to the court's belief that the individual realtor was on notice that insulation was in the house, and that she did not sufficiently investigate its composition. Nonetheless, the court refused to hold that the broker or the individual realtor had an obligation to investigate the nature of the insulation. The *Roberts* decision suggests that at least residential real estate brokers should strongly consider providing written information to prospective purchasers concerning the potential dangers arising from mold.

In addition, Pennsylvania courts have held that sellers have an affirmative duty to disclose known termite infestations, *see, e.g., Qhashnock v. Frost*, 445 A.2d 121 (Pa. Super. 1982), and other "serious and dangerous latent defects." *Anderson v. Harper*, 622 A.2d 319, 323 (Pa. Super. 1993). In *Anderson*, the Superior Court found that a seller of real property was obligated to disclose to his seller a malfunctioning sewage disposal system that caused sewage to surface to the top of the residential lot because it could pollute the environment, including water supplies, and might create potential health and safety problems. *Id.* at 324. The homeowner believed he had fixed the ongoing septic problem, but was deemed to know that the problem had not been fixed, and therefore had to be disclosed, because the seller did not obtain the permit required to fix the system and he did not properly fix it under existing regulations. *Id.* Since at present there are no regulations which specify what constitutes a complete remediation of mold conditions, good practice suggests that sellers obtain an independent third-party opinion of a qualified consultant that remediation has been completed before failing to disclose a pre-existing mold or moisture condition.

10. Negligent misrepresentation (sale of property)

To state a claim for negligent misrepresentation, the plaintiff must show: (1) misrepresentation of a material fact; (2) made under circumstances in which the misrepresenter ought to have know its falsity; (3) with an intent to induce another to act on it; and (4) which results in injury to a party acting in justifiable reliance on the misrepresentation. *Id.* at 561. Unlike intentional misrepresentation, the speaker need not know his or her words are untrue, but must have failed to make a reasonable investigation of their truth. In addition, there must be a duty owed by one party to another. *Id.* Without ruling on the issue, *Bortz* opined that real estate brokers might be subject to claims of liability under this theory, as well. *Id.* at 563. To date, no reported Pennsylvania case has found a duty to investigate conditions relating to mold. *See generally Gibbs v. Ernst*, 647 A.2d 882, 893 (Pa. 1994).

11. Innocent misrepresentation (sale of property)

An innocent misrepresentation is actionable if it is material to and the basis of a transaction, and the purchaser is justified in relying on the misrepresentation. *Id.* at 564. However, the remedy for an innocent misrepresentation is rescission, not monetary damages for tort recovery. *Id.*

F. Insurance coverage

There are not many reported decisions in Pennsylvania addressing insurance coverage for mold. One notable case is *Deede Smith Kemmerer v. State Farm Insurance Co.*, No. 01-5445, (E.D. Pa. Jan. 16, 2004), in which the plaintiff sought coverage for mold-related property damages allegedly attributable to a toilet leak at her home when she was on vacation. One of the covered perils was "sudden and accidental discharge or overflow," which the court stated would have justified recovery under the policy. However, the plaintiff's expert did not speak to the issue of causation. The insurance carrier submitted expert testimony that the property damage was caused by multiple uncovered causes of loss, including high humidity, stagnant air, a wet basement, dampness in the powder room from a neighboring greenhouse, and humid summer weather. The district court determined that the issues of causation in the case were complex and required expert testimony. Since the plaintiff had offered no such testimony, summary judgment was awarded to the defendant.

Chapter 10

Mold Litigation in Texas

Stephen Pate, Esq.

Synopsis

10.1 Introduction

Any recent article regarding mold litigation in the United States will invariably bring up the infamous *Ballard* case tried in Austin, Texas. In that 2001 case, a Texas jury handed down a $32.2 million verdict against the insurer of a house that was worth $750,000.[1,2] Though the Austin Court of Appeals reduced the award to $4 million plus interest and attorney's fees,[3] the verdict was the harbinger of a new era in mold litigation.

While *Ballard* was not the first "toxic mold" case, it attracted significant media attention with its enormous jury award. Media frenzy ensued and prompted the Commissioner of the Texas Department of Insurance to warn that fear and publicity were feeding upon each other to cause a mold dilemma.[4]

From 2000-2001, around the time *Ballard* was decided, insurance claims for mold damage in Texas increased exponentially from 1,050 claims in the first quarter of 2000 to 14,706 claims in the fourth quarter of 2001.[5] The number of claims made per 1,000 policyholders in Texas rose from 1.7 in the first quarter of 2000 to 23.57 in the fourth quarter of 2001.[6] By 2002, Texas, with 8 percent of the national population, had 75 percent of all mold claims.[7]

Insurance companies in the Texas homeowners insurance market were caught off guard by the surge in claims and the resulting hemorrhage of pay-outs.[8] In 2001, many of the larger insurers, including Allstate, State Farm, and Farmers Insurance Group, refused to issue new policies covering water-related claims.[9] By 2002, Farmers Insurance Group announced that it would not renew Texas homeowners' policies and several other insurers threatened to withdraw from the Texas homeowner's market. These moves were suspended only by agreements with the Texas Department of Insurance.[10]

At about the same time, insurance companies raised their rates. The average cost of insuring a home in Texas went from $786 a year in 2000 to $1,201 a year in 2003. In comparison, the nationwide average cost of insuring a home went from $508 a year to $569 a year during the same period of time.[11] The insurance companies attributed the higher rates in Texas to unwarranted lawsuits and higher claims for mold and storm related losses.[12] The dramatic increase in mold claims corresponds directly with an increase in the number of mold-related lawsuits, against insurance companies in particular.

This chapter outlines the issues related to mold litigation in Texas and summarizes a recent Texas statute that could affect mold litigation. Particular attention is paid to issues regarding accrual of action, admission of expert testimony in physical injury claims, and, in the insurance law arena, issues regarding trigger of policy coverage and ensuing loss clauses.

10.2 Mold Litigation in Texas

In mold-related lawsuits, plaintiffs have available to them numerous causes of action against a vast group of potential litigants. For example, if a plaintiff alleges that physical injury or property damage was caused by a defect in building construction, potential defendants could include builders, contractors, subcontractors, construction managers, property managers, architects, designers, construction component suppliers, and building owners. If a plaintiff alleges that mold damage was caused by faulty repair of household appliances or plumbing, the service person could be liable for the resulting damage, including property damage or physical injury. Some causes of action available to a plaintiff include breach of contract, breach of warranty, negligence, gross negligence, negligent misrepresentation, intentional or negligent infliction of emotional distress, and violations of the Texas Deceptive Trade Practices Act (DTPA).

As described earlier, homeowner insurers are likely the most often targeted defendants. Damages that may be claimed against insurers include the cost of mold remediation, compensation for property damage, physical injury, and the insurer's misconduct, if any, in handling the insured's claim. An insurer's misconduct in claims denial cascades beyond contractual claims involving the insurance policy and includes extracontractual claims arising from the denial of coverage. In such cases, a plaintiff has available, in addition to common law torts, Texas Insurance Code Article 21.21 actions for bad faith, unfair settlement practices, and unfair or deceptive acts at his disposal for causes of actions against an insurance company.

This subsection discusses the issues of when causes of action accrue for purposes of limitations, the difficulties in bringing claims for physical injury because of the difficulty in submitting expert opinions, and particular issues in claims against insurance companies, including statutory claims, coverage triggers, and ensuing loss clauses.

10.3 The Statute of Limitation and Accrual of Action Issues

Determining when a cause of action accrues, or for purposes of insurance coverage, determining when a policy is triggered, for mold-related damage is a challenging issue in mold litigation. Mold often grows in places where it is not easily discoverable by the building occupants. To deal with the particular challenges and circumstances related to injuries not easily discoverable, the Texas legislature has enacted statutes, and the courts have developed common law, to determine when an action should accrue.

For certain causes of action, the Texas Civil Practices and Remedies Code (CPRC) provides clear guidance regarding the period of limitations. Under the CPRC, most causes of action for personal injury and property damage have a two-year limitations period.[13] There is, however, an extended limitations period for improvement to building construction that may not be discoverable until many years after they were created.

For suits claiming personal injury or property damage against architects, engineers who furnish design, planning or inspection of construction of improvements to real property, and suits against persons furnishing construction or repair of improvements, suits may be brought within ten years after the substantial completion of the improvement.[14] This ten year limitations period applies regardless of when the defect is discovered.[15] The actions against these statutory defendants accrue when the improvements are substantially completed.[16]

Although the CPRC provides clear guidance as to accrual of action for claims against designers and builders of real property improvements, accrual of action for claims against other potential defendants are not so neatly resolved. Because mold damage may not be apparent until years after occurrence of the act or event that caused the mold to proliferate, determining when a cause of action accrues for the usual two-year limitations period is crucial.

Generally, a cause of action accrues when a wrongful act causes some legal injury, regardless of when the plaintiff learns of the injury.[17] A party needs only to be aware of enough facts to make him cognizant of his right to seek judicial remedy.[18] For mold-related damage, however, some questions have arisen regarding how to determine when the legal injury occurs because injury resulting from the wrongful act may not be readily apparent at the time the wrongful act leading to mold damage was committed.

To combat some of the uncertainties associated with the accrual of action when the injury is not readily apparent, Texas courts have formulated what is known as the "discovery rule," which is applied when the damage or injury was "inherently undiscoverable" and when the evidence of the damage or injury is objectively verifiable.[19] The discovery rule is a judicially conceived exception to the statute of limitations and is used to determine when a cause of action accrues where a plaintiff was unable to know of his injury at the time it occurred.[20] When applicable, the discovery rule tolls the running of the limitations period and defers the accrual of a cause of action until a plaintiff discovers, or through the exercise of reasonable care and diligence should discover, the nature of his injury.[21] The discovery rule imposes upon the plaintiff a duty to exercise reasonable diligence to discover facts of negligence or omission. Limitations period thus begins when the plaintiff knew or should have known of the facts giving rise to his cause of action.

Application of the discovery rule, however, is not always straightforward. Even when the discovery rule applies, it is still a question of fact when the plaintiff should have known the facts giving rise to his injury. *Booker v. Real Homes, Inc.* and *Palmer v. Sears, Roebuck & Co.* offer guidance on identifying the events triggering the accrual of a cause of action.

In *Booker*, the plaintiffs (the Bookers) first experienced water leakage problems due to construction defects in September 1997.[22] By November 1997, the Bookers began to notice a "musty" odor inside their home. They reported this odor to the builder of the house but the builder did not take any remedial action until the Bookers again asked the builder to remedy the problem in June 1998. In September 1998, the builder discovered water damage, including moisture and mold, and determined that the windows were the source of the water leaks. The builder and the window installer repaired the problem in November 1998 but the odor returned. In July of 1999, the Bookers discovered wet and rotten wood. The seepage from the window had apparently caused extensive water damage inside the walls of the house and the water damage led to the eventual infestation of mold. The mold, in turn, as the Bookers alleged, caused them health problems.

In October 1999, the Bookers sued the home builder and the window installer, claiming breach of contract, breach of warranty, negligence, gross negligence, negligent misrepresentation, intentional infliction of emotional distress, and violations of the Texas DTPA.[23] The trial court found, and the appellate court affirmed, however, that the Bookers' claims were barred by the two-year statute of limitation. In affirming the defendants' summary judgment, the appellate court applied the discovery rule and found that the limitations had run. That court explained that all that was required to commence the running of the limitations period was the discovery of an injury and its general cause, not the exact cause and fact and the specific parties responsible.[24] Thus, the court found that the discovery rule applied to toll the running of the statute until September 1997, the date on which the Bookers actually knew of the leaks, not when the Bookers determined the specific cause of the leaks and held that the period of limitations had run on the Bookers.[25]

A different situation was presented in *Palmer v. Sears, Roebuck & Co.*[26] In *Palmer*, Sears performed repairs on Palmer's air conditioner in April 1993. One month later, Palmer discovered a leak in the air conditioning unit where repairs were made that caused water to accumulate in the unit. In October 1994, Palmer became ill. A preliminary determination was made that mold and sporing from the air conditioner leak was the cause of Palmer's injuries. Palmer filed suit in September 1996.

Palmer acknowledged that he knew of his property damage, the first injury for which he could sue, in May 1993, more than two years before he filed suit. He insisted, however, that he could not have known of his physical injuries until they first manifested themselves in October of 1994, within the two-year limitations period.

The court, however, found that Palmer's cause of action accrued at the time Sears negligently performed the repairs to his air conditioner, and not on the date when the extent of the resulting damages were fully ascertainable.[27] Palmer's legal rights, property and physical, were invaded the moment water began leaking from the supposedly repaired air conditioner and damaged his property. In so holding, the court concluded that

> once injury results there is but a single tort and not a series of separate torts, one for each resultant harm. The cause of action thus created is for all the damage caused by a single legal wrong, and a plaintiff may not split this cause of action by seeking damages for some of his injuries in one suit and for later-developing injuries in another. The cause of action inheres in the causative aspects of a breach of a legal duty, the wrongful act itself, and not in the various forms of harm which result there from . . .[28]

Thus, under the discovery rule, a cause of action accrues when the plaintiff actually knows of the injury. It is not necessary for the plaintiff to determine the specific cause of the injury. Further, under the discovery rule, a cause of action accrues when the initial injury is discovered, not when subsequent injuries arise.

10.4 Difficulties in Maintaining Personal Injury Claims

Plaintiffs in mold-related litigation often claim physical injury resulting from occupying a mold-infested building. There are, however, particular difficulties in maintaining physical injury claims when the injuries are allegedly the result of exposure to mold. These difficulties include the plaintiff's burden to prove both general and specific causation. In other words, a plaintiff must prove that mold is capable of causing a particular injury or condition in the general public and that mold to which the plaintiff was exposed caused that plaintiff's injury, respectively.[29,30] Proving both types of causation has shown to be a difficult hurdle for the plaintiffs to overcome in Texas. In particular, general causation has been problematic to prove because few studies have been done on the effect of mold on physical health.

A. General requirements

The causation hurdle has proved to be difficult to overcome in Texas because Texas has adopted the *Daubert* standard for expert testimony in *Robinson v. E.I. Du Pont de Nemours and Co.*[31] Under *Robinson*, the trial judge acts as the gate-keeper of expert testimony evidence. The expert testimony must be sufficiently relevant and reliable to be admitted.

To prove general causation, a plaintiff generally must submit scientific evidence, including epidemiological studies, to show that mold is capable of causing the particular type of injury from which plaintiff suffers. Both *Daubert* and *Robinson* provide some factors in guiding the Texas courts in evaluating whether the proposed scientific evidence is reliable.[32] The *Daubert* and *Robinson* reliability requirement for scientific evidence of physical injury from mold is difficult to overcome because there have been few scientific studies done on the subject. Further, while epidemiological studies are generally used by experts where scientific evidence showing proof of causation is lacking, the Texas Supreme Court has mandated specific requirements for admission of epidemiological studies. [33]

B. Expert testimony generally

In Texas, for an expert witness's testimony to be admitted under the Texas Rule of Evidence, the expert must be qualified, the proposed testimony must be scientific knowledge (i.e. reliable), and the proposed testimony must assist the trier of fact to understand the evidence or to determine a fact in issue (i.e. relcvant).[34] Of these three requirements, reliability of the expert testimony is the most challenged requirement for cases relating to physical injury claims in mold-related litigation.

In Texas, reliability of an expert's testimony may be challenged on four grounds—the foundational facts of the expert's opinion, the methodology the expert employed, whether there is an "analytical gap" between the expert's conclusion and the underlying data, and whether the expert accounted for other plausible causes of the damage.[35]

Further, the Texas Supreme Court enumerated a nonexclusive list of six factors to assist the trial court in determining the reliability of expert testimony: (1) extent to which the theory has or can be tested; (2) extent to which the theory relies on expert interpretation; (3) whether the theory has been subjected to peer review; (4) the technique's potential rate of error; (5) the technique's acceptance by the relevant scientific community; and (6) the technique's non-judicial uses.[36]

1. Few studies have been done to shown the causal link between mold toxins and physical injury

The presence of mold, by itself, is not believed to necessarily cause harmful effects to human health. It is the toxin that mold releases during sporing, or mycotoxins, that is the suspected culprit behind human ailments some believe to result from exposure to mold. Thus, the presence of mold in a building does not necessarily mean the occupants will be exposed to mycotoxins. For a person to be exposed to the mycotoxins, mold spores must be released into the air and inhaled, contacted, or ingested. Virtually all information related to diseases caused by mycotoxins concern ingestion of foods contaminated with mold.[37] There has been a paucity of studies done to examine the exposure to mycotoxins through inhalation.

Although epidemiological studies have been done and those studies have consistently detected an association between respiratory symptoms and mold growth in homes, causality in those studies has not been established.[38] In September 2002, the Texas Medical Association (TMA) adopted a report from its counsel which concluded that adverse health effects from exposure to *Stachybotrys* spores (spores from the *Stachybotrys chartarum* strain of mold) were not supported by available peer review reports in medical literature.[39] The TMA instead concluded that the "probability or possibility" of causation or exacerbation of medical conditions due to exposure to mold in an indoor environment was limited to those individuals who have respiratory allergies to *Stachybotrys* spores or those rare individuals who have rare inflammatory lung diseases.[40] The author of that report acknowledged that there was a "fundamental disconnect" between what the public was being told by the media and consensus within the scientific community.[41]

Consistent with the TMA's finding, the US Environmental Protection Agency has not set standards or Threshold Limit Values (TLVs) for airborne concentrations of mold, or mold spores. Currently there are no EPA regulations or standards for airborne mold contaminants.[42] The EPA thus far only recognizes that molds can trigger asthma episodes in sensitive individuals with asthma and trigger allergies in sensitive individuals.[43]

A recent study conducted by the Institute of Medicine, on behalf of the U.S. Centers for Disease Control and Prevention, concluded that toxins from mold and bacteria can cause coughing and wheezing, but that there is no evidence linking mold to more serious ailments such as cancer, fatigue or neurological problems.[44] Even the study's conclusion linking the association between mold and coughing and wheezing must be considered with prudence because the study did not specify which agents were present in the damp buildings occupied by the research subjects.[45] Bacterial toxins could very well have been in the buildings, and in fact were expected to be in the buildings, in addition to mycotoxins from molds. While the study found that it is *biologically plausible* that toxins from *Stachybotrys chartarum* could cause effects in humans, the study cautiously urges validation from more extensive research before conclusions are drawn.

2. Case law regarding admission of expert testimony on the effects of mycotoxin on human health

The lack of reliable scientific studies demonstrating the link between mycotoxins and human health effects makes it difficult for a plaintiff to admit expert testimony regarding causation. Cases in Texas generally reflect this difficulty.

In *Flores v. Allstate Texas Lloyds Co.*, a family (the Flores) sued their homeowner's insurance company, claiming that the mold in the home caused allergic health effects.[46] The Flores attempted to designate their family physician as a causation expert on the health effects of the household mold. The court examined the proposed expert testimony against the four *Daubert* factors. The court first concluded that plaintiffs' theory that their allergies were caused by mold in their homes was not tested by their designated expert or any other expert. Second, that the expert's proffered testimony failed the "peer review" inquiry of *Daubert* because he was not able to point to publication or study that supported his hypothesis that household molds cause allergies. Third, the potential for error in the proposed expert's technique is "virtually limitless" because the expert did not conduct tests on the plaintiffs nor was the expert able to identify any scientific study that demonstrate that the type of mold found in plaintiffs' home more likely than not caused them any health effects. Finally, the court found that the expert failed to present evidence as to the acceptance of his theory in the scientific community. Based on this four-part *Daubert* analysis, the court held that the Flores failed to prove general causation.

Further, the court found there was no specific causation because no one in the plaintiffs' household had undergone any testing to determine allergies to household mold and the expert failed to account for other factors that may have caused the plaintiffs' health problems. For example, cigarettes, dust mites, fibers, mesquite, or other environmental allergens. Specifically, the expert never conducted allergy tests on the plaintiffs or evaluated plaintiffs' exposure to mold in their home. The court thus excluded the expert's testimony.

The *Ballard* case is perhaps the most instructive case showcasing the high hurdle that potential expert witnesses must overcome before an expert can be found sufficiently reliable to testify on a subject matter. In *Ballard*, the plaintiff's husband, Allison, was diagnosed with a type of brain damage called toxic encephalopathy.[47] Two leading experts in the study of health effects of mold and mycotoxins testified on behalf of Allison and agreed that exposure to mold caused his toxic encephalopathy. The trial court, however, found that even though the foundational data used by the experts to prove general causation met the requirements of *Daubert/Robinson*, it was unreliable according to the factors propounded by the Texas Supreme Court in *Havner v. Merrill Dow Pharmaceuticals, Inc.*[48]

In *Ballard*, Allison's causation experts had presented an epidemiological study which the court found to be a "crucial underpinning" of their opinion. The court read *Havner* to hold that for an epidemiological study to be reliable foundation, it must be (1) unbiased in its design and show that exposure to the allegedly toxic substance more than doubles the risk of injury and (2) it must be capable of repetition with the same results 95 percent of the time (a confidence interval of 95 percent). The court found both elements lacking in the epidemiological study relied upon by Allison's experts. The experts had failed to show that exposure to mold more than doubled the risk of injury and further could not say whether their technique was generally accepted. Allison's experts also could not provide a confidence interval because there had not been sufficient studies done. One of Allison's experts admitted that "there are very few case reports that toxic molds inside homes can cause unique or rare health conditions. . .. These reports are rare, and a causal link between the presence of toxic mold and these conditions has not been proven."

Even though the *Ballard* court seems to suggest that the requirements for 95 percent confidence interval and proof that exposure to mold more than doubles the risk of injury are necessary only if the expert relies on epidemiological studies, it found that the study was a "crucial underpinning" of the experts' opinions. At this time, it is not yet clear whether potential experts will be able to demonstrate reliability of their testimony without presenting epidemiological studies. If so, plaintiffs may be able to bypass the stringent requirements that *Havner* appears to impose on epidemiological studies before the studies could be admitted. It remains likely, however, that epidemiological studies will continue to be crucial to mold experts' opinions because no studies have shown a direct causal link between mold and physical injuries.

A possible counterexample to the stringent reliability requirements is the more recent case of *Caldwell v. Curioni*.[49] In *Caldwell*, the plaintiffs (Caldwells) became ill soon after they moved into a rented house. An examination of the house revealed a large amount of mold.

The Caldwells had offered as evidence a report by one of their experts which confirmed the presence of *Stachybotrys*. The report concluded that *Stachybotrys* produces a mycotoxin that has been documented as toxic to humans. Further, the Caldwells offered affidavits of doctors they had consulted who testified the basis for their opinions and stated that they observed a multitude of physical problems in the Caldwells which were caused by mold infestation. The court appeared to have accepted the experts' opinion that the Caldwells physical injuries were caused by mold.

The court did not specifically go through the *Robinson* and *Havner* analysis and the defendant did not appear to have challenged the reliability of Caldwells' expert testimony. The court did appear to accept that the Caldwells provided at least some evidence of general and specific causation and the court found that the defendant's failure to eliminate the mold before the Caldwells occupied the house was the proximate cause of their physical injuries.

The *Caldwell* case appears to demonstrate that courts are at least willing to consider expert testimony of the adverse health effects of mycotoxins without undergoing the rigorous analysis under *Robinson* and *Havner*. However, the lack of scientific evidence that mold causes serious health problems probably dooms these efforts.

10.5 Insurance Litigation

As noted earlier, no industry is affected greater by the rise in mold litigation than insurance companies, who are the targets of most of the litigation resulting from mold damage. Not only are the insurance companies liable for first party claims under homeowner insurance policies, they are often also liable for third party claims under comprehensive general liability policies issued to builders or contractors.

Texas has one of the most heavily regulated insurance industries in the country. Until 2003, only two basic forms for the homeowner's insurance policy were allowed: HO-A and the more comprehensive and more popular, HO-B.[50] Since the insurers threatened to withdraw from the Texas homeowner's insurance market and to deal with the skyrocketing rates, the Department of Insurance has permitted more forms that provide for more explicit exclusions and inclusions of claims regarding mold and additional endorsements for mold coverage.[51]

While much controversy surrounds what is covered under a policy, significant mold verdicts against insurers almost always involve bad faith or poor claims-handling issues. For example, in *Ballard*, the defense attorney referred to the case as a mold case but the jurors pointed out that it was a bad faith case.[52] It is the breach of extracontractual duties that leads to the large awards that have grabbed the headlines. In addition to common law causes of action, the Texas Insurance Code provides stiff penalties against the insurers for misconduct against the insured.

This section briefly discusses some of the possible statutory claims under the Texas Insurance Code and two particular issues that have arisen in regard to litigation involving insurance companies—the trigger event of an occurrence policy and the "ensuing loss" provision common in homeowner insurance policies. These two areas represent two unsettled areas of insurance law in Texas regarding mold-related litigation.

A. Statutory claims under the Texas Insurance Code Articles 21.21 and 21.55

Article 21.21 of the Texas Insurance Code is the primary source of law for bad faith and misrepresentation claims against insurers in Texas. Under Article 21.21, insurers are prohibited from unfair or deceptive trade practices.[53] Section 4(10) of the Article provides a laundry list of ten prohibited acts and practices which the Article defines to be unfair settlement practices.

For example, Section 4(10)(a)(ii) prohibits insurers from failing to attempt in good faith to effectuate a prompt, fair, and equitable settlement of a claim with respect to which the insurer's liability has become reasonably clear. This is the standard "bad faith" claim that insured's bring against their insurers for failing to pay a claim and the standard is identical to the common law bad faith standard.[54]

The statutory claim, however, provides for an award of up to three times actual damages if the insurer's wrongful conduct is committed "knowingly."[55] The prevailing claimant may also recover court costs and reasonable attorney fees.[56] Further, Article 21.21 provides that if a plaintiff can show that he relied on an insurer's misconduct to his detriment, the plaintiff can bring an action under the Texas Deceptive Trade Practices Act, provided that other prerequisites for bringing an action under the DTPA are met.[57]

B. Trigger of insurance coverage

A significant point of contention is often in determining the trigger event of an occurrence policy. Most homeowner insurance policies are occurrence policies for which coverage is triggered by the occurrence of a covered event. Similar to the problem associated with accrual of action, a difficulty exists in determining when mold damage "occurred" and when coverage is triggered.

For first party property damage claims, some Texas courts have adopted the "relaxed manifestation" theory to designate the trigger of an occurrence policy.[58,59] Under the relaxed manifestation theory, a loss or damage occurs when it manifests or when it becomes readily apparent, not when the wrongful act was committed.[60] A loss or damage is considered apparent when it is "capable of being easily perceived, recognized and understood."[61] This theory has a striking resemblance to the discovery rule for deferral of the accrual of causes of action.

American Home Insurance Co. v. Unitramp illustrates the application of the manifestation theory in Texas.[62] In *Unitramp*, the claimed injury was the receipt of fuel diluted with water onto a ship. The issue was whether the injury to the purchaser of the fuel manifested when the watered-down fuel was received (which was prior to the relevant policy period) or when the watered-down nature of the fuel became apparent (which was during the policy period). Because no one knew the fuel was diluted when it was loaded onto the ship, the court held that the resulting damage did not manifest until tests revealed that the fuel was diluted. *Id.* Prior to the testing of the fuel, the damage to the purchaser was not readily apparent or easily perceivable.

Similarly, in mold-related cases, some courts must hold that the presence of mold would most likely be considered to trigger coverage in an occurrence policy only when it is manifested or becomes readily apparent, not when it is actually discovered or when the causal act was committed. The benefit of the relaxed manifestation theory some commentators believe lies in the balance of providing the insured's relief from having to discover harm when it is not discoverable yet putting the insured's on notice of their duty of diligence in discovering the harm. It should be noted that it is unclear what standard will be adopted in first party cases.

For a comprehensive general liability (CGL) policy, however, there is authority that the exposure theory is used to determine trigger of coverage where a triggering occurrence includes continuous or repeated exposure to conditions under the policy.[63] Under the exposure theory, coverage is triggered in any policy period in which *exposure* to cause of injury occurred.[64] Time of discovery or manifestation of damage is irrelevant. Further, while the manifestation theory is applied only to property damage claims under first party policies, for a CGL policy, the exposure theory is applied to both physical injury and property damage.[65]

C. "Ensuing loss" provisions

One of the most unsettled areas of Texas insurance law regarding mold damage is the "ensuing loss" provision in insurance policies. A typical policy contains a clause that expressly excludes certain perils, for example:

> Loss caused by wear and tear; deterioration; rust, rot, mold or other fungi; dampness of atmosphere, extremes of temperature; contamination; vermin, termites, moths or other insects;

This exclusionary clause is usually modified by an "ensuing loss" provision, for example:

> The foregoing Exclusions shall not apply to ensuing loss caused by collapse of building, or any part thereof, water damage or breakage of glass which constitute a part of the building, provided such losses would otherwise be covered under this policy.[66]

On its face, the ensuing loss provision does not appear to provide coverage for mold loss. As an illustration, the above sample provisions expressly excludes loss from mold. The ensuing loss provision thus reads that mold loss would not be excluded from policy coverage if the mold loss is a loss that ensues from, for example, water damage, *if* the mold loss would otherwise be covered under the policy. Mold loss, however, is not otherwise covered under the policy because it is expressly excluded by the exclusionary clause. Thus, the ensuing loss provision would not provide coverage for mold loss. Under this interpretation, the ensuing loss provision does not alter the exclusion against mold damage.

This is most clearly illustrated in the unreported case *Fiess v. State Farm Lloyds*, a decision by the Southern District of Texas. In *Fiess*, Tropical Storm Allison flooded the Fiesses' house and, in turn, caused mold to proliferate in the house.[67] Although mold was expressly excluded by the Fiesses' homeowner's policy, the Fiesses argued that the mold exclusion did not apply because their mold damage falls within an exception to the exclusion as an ensuing loss caused by water damage under the policy. The Southern District court, however, held that the ensuing loss provision did not work to except the mold damage from the express policy exclusion because the mold was caused by water damage.

The court reasoned that "ensuing loss" referred to the loss that was a consequence of an excluded loss, not loss that caused an excluded loss.[68] Under this reasoning, loss caused by mold would not be covered. However, if mold caused water damage (or any of the other enumerated losses in the ensuing loss provision), the loss that results from the water damage, that was itself caused by mold, would be covered. In other words, if mold causes water damage, which in turn causes loss, the loss is the "ensuing loss" that is covered by the provision. The loss, which is ultimately caused by mold, would be covered, despite the exclusion of mold damage.

This view was grounded on a prior Texas case which dealt with settlement of foundation. In *Lambros v. The Standard Fire Insurance Co.*, subsurface water in the plaintiff's home caused the building's foundation to settle, which in turn caused the plaintiff's loss.[69] The plaintiff's policy excluded loss caused by settling of foundation but contained an ensuing loss clause that excepted losses caused by water damage from the exclusions. The court found no coverage because the settlement of foundation was caused by water damage and not vice versa. Giving the ensuing loss exception its "ordinary meaning," the court concluded that "an ensuing loss caused by water damage is a loss caused by water damage where the water damage itself is the result of a preceding cause." [70] Thus, the ensuing loss exception comes into effect when an excluded loss causes water damage, which in turn causes a loss. In other words, the ensuing loss exception would not cover an otherwise excluded loss unless the otherwise excluded loss caused one of the enumerated damages in the ensuing loss provision which in turn caused the claimed loss.

Lambros is particularly instructive because the Texas Supreme Court has refused writ on the case, meaning that the Texas Supreme Court implicitly found the judgment of the court of appeals, and the legal principles announced in the opinion, to be likely correct.[71] This theoretically elevates the opinion of the appeals court to having the same precedential value as an opinion of the Texas Supreme Court.[72]

Despite *Lambros*, a divergent line of cases in Texas courts have, however, interpreted the ensuing loss provision to cover mold loss. Under this second line of cases, the courts have held that the ensuing loss provision provides an exception to an express exclusion if the otherwise excluded loss results from one of the enumerated causes in the ensuing loss provision. Under this theory, for example, mold loss would be covered, despite its express exclusion, if it is the result of water damage.

A recent opinion in the Southern District of Texas, McAllen Division, announced one month after the *Fiess* decision, espouses this second interpretation. In *Flores v. Allstate Texas Lloyd's Company*, the court held that the mold exclusion precluded coverage for mold occurring naturally or from non-covered events, but did not preclude coverage for mold "ensuing" from a covered water damage event.[73] In other words, the ensuing loss provision provides coverage for an otherwise excluded loss when the otherwise excluded loss was caused by a covered event, regardless of the cause of the covered event.

Similarly, the Southern District of Texas, Houston Division, held in *Coury v. Allstate Texas Lloyds* (March 31, 2004) that although mold was generally excluded by the policy, mold may be covered by the exclusion's "ensuing loss" language.[74] The court held this policy language should not be read as always excluding mold coverage. The court found that the phrase "if the loss would otherwise be covered under this policy" applied to the water damage event, not the mold damage event. Under this interpretation, the "loss" is the water damage, not the excluded mold. The court expressly declined to follow *Fiess* because that holding required "manipulation of the [policy] language." The court found its reasoning consistent with the result of the Fifth Circuit opinion in *Aetna Cas. & Sur. Co. v. Yates* because although *Yates* held that mold was not a covered ensuing loss, the *Yates* court opined that mold might have been covered under different facts.[75] The court found that in *Yates*, the "otherwise covered loss" was alleged to be water damage and that the damage that ensued from that loss was alleged to be mold.

Two months after the *Coury* decision, however, the same court granted Summary Judgment against an insured in *Hood v. State Farm Lloyds* and held that no matter how mold is caused, it is not covered because mold damage is specifically excluded from coverage in the policy.[76,77] The *Hood* court cited *Fiess* as authority.

It is unclear how Texas courts would resolve the inconsistent holdings in the two lines of cases. Even the same court in the Southern District cannot agree and propounded different interpretation of virtually identical ensuing loss provisions. The Fifth Circuit certified the *Fiess* appeal to the

Texas Supreme Court. That court held arguments on March 31, 2005 and a decision is pending.

10.6 Texas Mold Legislation

The increase in public awareness (or, in some parties' view, unmitigated fear) of the effects of mold has caused a corresponding increase in the number of industrial hygienists and mold remediators. Industrial hygienists inspect buildings for presence of mold. Remediators remove and clean building structures, materials, and personal belongings contaminated with mold. In Houston alone, there are approximately 300 businesses that offer remediation services. This cottage industry is formed by water extractors, asbestos removal experts, contractors, and, as some have termed, the "get rich quick off of mold" group.[78] The previous lack of standards and regulations in this area has caused some to worry about fraud on frightened consumers by unscrupulous remediators or inspectors.

In 2003, the Texas legislature sought to address this issue and enacted House Bill No. 329, which provides for rules regulating mold assessors and remediators, clarifies liability of property owners, and prohibits insurer discrimination against buildings previously contaminated by mold.[79]

Bill 329 added Chapter 1958 to the Texas Occupations Code, prohibiting individuals from inspecting or investigating buildings for the presence of mold unless the individual has a mold assessment license and prohibiting individuals from mold remediation without a mold remediation license. By January 1, 2005, all remediators and inspectors must obtain the proper licenses to continue to conduct mold-related activities. To obtain a license, an individual must successfully complete a training program offered by a Texas Department of Health accredited trainer and must successfully pass a licensure examination. Pursuant to Chapter 1958, the Texas Board of Health has adopted the Mold Assessment and Remediation Rules that specifically lay out how remediators and inspectors are regulated, including a Code of Ethics.[80]

The Bill seeks to protect consumers by requiring mold assessors to provide a standardized mold assessment report and an evaluation of the mold remediation project to certify that the mold will not return from the remediated cause. The mold remediator also must prepare a "remediation work plan" to the consumer before mold remediation begins.

Importantly, the Bill exempts property owners from civil liability derived from damages related to a decision to allow someone to occupy a building after mold remediation has been performed on the property.

Finally, House Bill 329 amended the Texas Insurance Code by adding Article 21.21-11. The new subchapter affects all insurers that write residential property insurance in Texas. This section prohibits insurers from making underwriting decisions regarding residential property insurance policies based on previous mold damage or claim for mold damage. If the insurance applicant has property eligible for homeowner insurance but has had mold damage, the insurer cannot discriminate against the applicant in issuing insurance policies if the property has been remediated and current inspection reveals that there is no evidence of mold damage.

Endnotes

1. *Ballard v. Fire Ins. Exch.*, No. 99-05252 (250th Dist. Ct., Travis County, Tex., June 1, 2001).

2. *See Allison v. Fire Ins. Exchange*, 98 S.W.3d 227 (Tex. App.-Austin 2002, pet. denied).

3. *Id.*

4. Texas Department of Insurance, Commissioner Montemayor Statement to Consumers and Consumer Groups, Sept. 18, 2001, at http://www.tdi.state.tx.us/commish/moldconsumer.html.

5. Texas Department of Insurance, *Mold Data-All Claims*, June 14, 2002, at http://www.tdi.state.tx.us/commish/news/molddata2.html.

6. *Id.*

7. Mary Sit-DuVall, "Claims Boost Insurance Rates, Fraud," *Houston Chronicle*, Aug. 30, 2002, at http://www.chron.com/cs/CDA/story.hts/topstory2/1475656.

8. *Id.*

9. *See, e.g.*, Terrence Stutz, "Allstate Halts New Policies: Insurer is Latest in Texas to Cite Mold Losses in Homeowner's Market," *Dallas Morning News*, Sept. 29, 2001, at 1A, available at 2001 WL 28629932.

10. *The Washington Post*, Nov. 12, 2002, at E02, 2002 WL 102571227.

11. Christopher Clausen, "Texas consumers fight high cost of home insurance," *The Beaumont Enterprise*, Apr. 23, 2004.

12. *Id.*

13. Civ. Prac. & Rem. Code § 16.003 (2004).

14. Civ. Prac. & Rem. Code §§ 16.008-16.009 (2004).

15. *See Trinity River Auth. v. URS Consultants, Inc.*, 889 S.W.2d 259, 260 (Tex. 1994).

16. See note 14.

17. *See Moreno v. Sterling Drug, Inc.*, 787 S.W.2d 348, 351 (Tex. 1990).

18. *See Murray v. San Jacinto Agency, Inc.*, 800 S.W.2d 826, 828 (Tex. 1990).

19. *See Velsicol Chem. Corp. v. Winograd*, 956 S.W.2d 529, 531 (Tex. 1997).

20. See Moreno, 787 S.W.2d at 351.

21. *See id; HECI Explor. Co. v. Neel*, 982 S.W.2d 881, 886 (Tex. 1998).

22. *Booker v. Real Homes, Inc.*, 103 S.W.3d 487 (Tex. App.-San Antonio 2003).

23. *See id.* at 490.

24. *See id.* at 492.

25. The court, however, found that there was evidence of fraudulent concealment on the part of the builder and reversed the summary judgment in favor of the builders. Fraudulent concealment estops a defendant from using limitations as an affirmative defense and suspends the running of the limitations period after it has begun. *See Booker*, 103 S.W.3d at 493.

26. *Palmer v. Sears, Roebuck & Co.*, 969 S.W.2d 582 (Tex. App.-Fort Worth 1998).

27. Though the court held that the action accrued when the repair was made, subsequent language suggests that the action accrued when Palmer detected the leakage, one month after the repair was made. Having the action accrue when Palmer detected the leakage would be in line with the court's reasoning.

28. *Palmer v. Sears, Roebuck & Co.*, 969 S.W.2d at 585, *quoting Gideon v. Johns-Manville Sales Corp.*, 761 F.2d 1129, 1136-37 (5th Cir. 1985).

29. *Merrell Dow Pharm., Inc. v. Havner*, 953 S.W.2d 706, 714 (Tex. 1996).

30. *Havner*, 953 S.W.2d at 714.

31. *See Robinson v. E.I. Du Pont de Nemours and Co.*, 923 S.W.2d 549 (Tex. 1995).

32. *Flores v. Allstate Texas Lloyd's Co.*, 229 F. Supp. 2d 697, 701 n.6 (5th Cir. 2002).

33. *See Havner*, 953 S.W.2d at 717.

34. *E.I. Du Pont de Nemours and Co. v. Robinson*, 923 S.W.2d at 556.

35. *Allstate Texas Lloyds v. Mason*, 123 S.W.3d 690 (Tex. App.— Fort Worth 2003, no pet. his.)

36. *Robinson*, 923 S.W.2d at 556.

37. U.S. Environmental Protection Agency, *Indoor Air Pollution: An Introduction for Health Professionals*, Mycotoxins, 1994, at http://www.epa.gov/iaq/pubs/hpguide.html.

38. James Guy, Risks and Liabilities Associated with Toxic Molds, Address at the 2001 Environmental Law Update Seminar (Oct. 2001).

39. O. Edwin McClusky, Report of Council on Scientific Affairs: Black Mold and Human Illness, available at http://www.texmed.org/has/csa black mold.doc.

40. *See id.*

41. *See id.*

42. http://www.epa.gov/iaq/molds/moldresources.html

43. *See id.*

44. Board on Health Promotion and Disease Prevention, Institute of Medicine, Damp Indoor Spaces and Health, National Academies Press, p. 6, 109-54 (2004), *prepublication copy available at* http://www.nap.edu/books/0309091934/html/.

45. *Id.*, at p.8.

46. *See Flores v. Allstate Texas Lloyd's Co.*, 229 F. Supp. 2d 697 (S.D. Tex. 2002)

47. *See Allison.*

48. *See Allison; see also Merrell Dow Pharm. Inc. v. Havner*, 953 S.W.2d 706 (Tex. 1997).

49. *See Caldwell v. Curioni*, 125 S.W.3d 784 (Tex. App.— Dallas 2004, no pet. hist.).

50. The Texas HO-B policy was problematic for insurers because it did not require water discharge to be sudden and accidental, as is required by standard insurance policies issued in most states. Krik Hansen, "Mold: The Fugus Among Us is Difficult to Evict," *Insurance Journal*, May 5, 2003, at http://www.insurancejournal.com/magazines/west/2003/05/05/partingshots/28748.htm.

51. The Texas Department of Insurance now allows more forms of homeowner insurance policies. (The policies in Texas now specifically exclude mold, can purchase mold endorsements separately.) Most policies do not cover mold remediation beyond that necessary to repair or replace property damaged by water loss otherwise covered by the policy. The HO-A policy offers no coverage for mold remediation or for damage caused by water leaks, although some companies offer coverage for sudden and accidental water leaks as an endorsement to the base HO-A policy. Insurance companies, however, are required to offer mold remediation coverage. The HO-B policy is much more expensive and covers mold damage only if the mold is on property physically damaged by sudden water and accidental water. This includes hidden and concealed damage that is reported within 30 days of detection. However, even the HO-B policy does not cover cost of testing or remediation. Additional endorsements would have to be purchased.

52. Fred Hagans and Jennifer Rustay, "Mold Bad Faith Cases: Tools and Strategies," *The Brief*, Fall 2003 at 13.

53. See Tex. Ins. Code Ann. Art. 21.21 (Vernon 2004).

54. *See Mid-Century Ins. Co. v. Boyte*, 80 S.W.3d 546, 549 (Tex. 2002).

55. *See supra* note 57 § 16(a).

56. *Id.*

57. *Id.*

58. The Texas Supreme Court described the theory prescribed by the Dallas Court of Appeals in *Dorchester Development Corp. v. Safeco Insurance Co.*, 737 S.W.2d 380, 383 (Tex. App.— Dallas 1987, no writ), as a "pure" manifestation approach. *See American Physicians Ins. Exchange v. Garcia*, 876 S.W.2d 842, 853 n.20 (Tex. 1994). The approach taken in *Dorchester*, however, was actually the "relaxed" manifestation approach, as a later Dallas Court of Appeals case clarified. *See Cullen/Frost Bank v. Commonwealth Lloyd's, Insurance Co.*, 852 S.W.2d 252, 257 (Tex. App.-- Dallas 1993, writ denied).

59. The "pure" manifestation rule triggers coverage upon actual discovery of injury whereas the "relaxed" manifestation rule triggers coverage in the first policy period during which discovery of injury is possible. *See American Physician Ins. Exch. v. Garcia*, 876 S.W.2d 842, 853 n.20 (Tex. 1994).

60. *American Home Assurance Co. v. Unitramp Ltd.*, 146 F.3d 311, 313 (5th Cir. 1998); *Dorchester Dev. Corp. v. Safeco Ins. Co.*, 737 S.W.2d 380, 383 (Tex. App. – Dallas 1987, no writ).

61. *Unitramp*, 146 F.3d at 314.

62. *American Home Ins. Co. v. Unitramp*, 146 F. 3d 311 (5th Cir. 1998).

63. *Pilgram Enterprises, Inc. v. Marland Cas. Co.*, 24 S.W.3d 488 (Tex. App.— Houston [1st Dist.] 2000, no pet.).

64. *American Physician Exch.*, 876 S.W.2d at 853 n.20.

65. *Pilgram Enterprises, Inc.*, 24 S.W.3d at 497-98.

66. *See, e.g., Merrimack Mut. Fire Ins. Co. v. McCaffree*, 486 S.W.2d 616 (Tex. Civ. App. – Dallas 1971, writ ref. n.r.e.) (emphasis added).

67. *Fiess v. State Farm Lloyds*, 2003 WL 21659408 (S.D. Tex. June 4, 2003).

68. *See id.* ("'ensuing loss caused by . . . water damage' refers to water damage which is the result, rather than the cause, of one of the types of damage enumerated in [the exclusion]"); *see also Lambros v. Standard Fire Ins. Co.*, 530 S.W.2d 138 (Tex. Civ. App.— San Antonio 1975, writ denied).

69. *Lambros v. Standard Fire Ins. Co.*, 530 S.W.2d 138 (Tex. Civ. App.— San Antonio 1975, writ denied).

70. *Id.* at 141.

71. Texas Rules of Form, Texas Law Review, Tenth Edition, Appendix A.

72. See *id.*

73. *Flores v. Allstate Texas Lloyd's Co.*, 278 F. Supp. 2d 810, 814 n.3 (S.D. Tex. 2003).

74. *Coury v. Allstate Texas Lloyds*, Cause No. H-02-2238 (S.D. Tex. March 31, 2004).

75. *See Aetna Cas. & Sur. Co. v. Yates*, 344 F.2d 939, 941 (5th Cir. 1965)

76. *Hood v. State Farm Lloyds*, Cause No. H-03-2621 (S.D. Tex. May 17, 2004).

77. The *Hood* opinion was issued by Judge Lynn Hughes while the *Coury* opinion was issued by Judge Nancy Johnson. Both judges sit in the Houston Division of the Southern District of Texas.

78. Mary Sit-DuVall, "The Mold Toll: Claims Boosts Homeowner Insurance Rates, Fraud Complaints," *Houston Chronicle*, Sun June 30, 2002, at A1.

79. Tex. H.B. 329, 78th Leg., R.S. (2003).

80. *Id.*

About the Editors

Joel L. Herz, Esquire of the Law Offices of Joel L. Herz in Tucson, Arizona concentrates on complex environmental and related litigation matters. He graduated from the University of Michigan with honors in 1985, and the University of Michigan Law School in 1987. From 1987 to 1994, Joel Herz practiced in the environmental and litigation practice groups with the New York office of Fried, Frank, Harris, Shriver and Jacobsen. He has been involved in numerous mold litigations. He also counsels and advises local developers, businesses and attorneys with regard to mold issues and the acquisition and sale of environmentally tainted properties. He is a contributing author of *Brownfields: A Comprehensive Guide To Redeveloping Contaminated Property* published by the American Bar Association Section of Natural Resources, Energy and Environmental Law (1997). Joel Herz is AV rated by Martindale-Hubbell.

Kimberly A. Taylor, J.D. graduated cum laude from Wellesley College and earned her Juris Doctor degree from the University of Arizona College of Law. After practicing law for several years she now devotes herself full-time to a career in freelance legal writing. Other publications include several chapters on legal ethics in the High Court Case Summaries law study guide series, as well as the forthcoming *Witness:Truth—An Introduction to Forensic Science for High School Students* (Textbook and Teachers Manual) from Lawyers and Judges Publishing Company, Inc.

About the Contributors

Barry L. Bellovin, Attorney and Counselor at Law, has been a shareholder in Bellovin & Karnas, P.C. since 1985 (formerly Siegel, Bellovin & Karnas P.C.) in Tucson, Arizona. His practices focuses extensively on plaintiff personal injury, medical malpractice, class action, product liability and toxic tort ligitation. Since 1995, Mr. Bellovin has served as Judge for the Juvenile Court Commissioner, Pima County Juvenile Court in Tucson. Mr. Bellovin has also served as Judge to the Special Pima County Superior Court Commissions and Special Magistrate in Tucson, and has taught law, business law and social science at the Pima Community College and University of Phoenix in Tucson.

Mr. Bellovin is a member of the State Bar of Arizona, California State Bar, New York State Bar, U.S. Supreme Court Bar, U.S. Court of Appeals (9th Circuit), U.S. District Court of Arizona, U.S. District Court of California (Northern, Eastern, Central and Southern Districts), U.S. District Court of New York (Southern and Eastern Districts), American Bar Association, Association of Trial Lawyers of America, Arizona Trial Lawyers, Pima County Bar Association, and the Hastings College of Law Alumni Association. He is also active in community service serving as president or board member for numerous organizations. In 1996, Mr. Bellovin received the Pima County Bar Association's Pro Bono Attorney Award.

Edward Eshoo, Jr. is a founding partner of Childress Duffy Goldblatt, Ltd. in Chicago, where his areas of practice include insurance coverage, insurance producer malpractice, mold litigation, commercial litigation and construction defects. Mr. Eshoo has lectured extensively on a variety of topics related to mold litigation, including seminars on "Mold Litigation from the

Policyholder's Perspective," "Making Your Insurance Company Pay for Toxic Mold," "Anatomy of a Mold Case," and "Challenging Mold-Related Exclusions and Limitations in Homeowners Insurance Policies." He is a member of the American, Illinois State, and Chicago Bar Associations, the National Association of Public Insurance Adjusters, the National Fire Protection Association, the Illinois Trial Lawyers Association, and the Association of Trial Lawyers of America, for whom he has co-chaired the Toxic Mold Litigation Group. Over the past nineteen years, he has lectured on insurance-related topics and has authored articles and publications on first-party property insurance issues, including chapters four and ten of the *Illinois Institute of Continuing Legal Education Property Insurance Practice Handbook.*

Michaela A. Fanning graduated from Boston College Law School in 1986 and served as a law clerk to Chief Justice Francis J. Boyle, United States District Court, District of Rhode Island. She is admitted to practice law in Massachusetts and Rhode Island and has litigation experience in state and federal trial courts, as well as at the appellate level.

Ms. Fanning focuses her practice on first-party insurance contract coverage disputes, and third-party defense of personal injury claims, particularly claims involving lead, mold and other toxic exposures. She served as co-counsel in the trials of four lead-paint poisoning cases in the Massachusetts Superior and Housing Courts. Two of the cases resulted in defense verdicts, one resulted in a plaintiff's verdict of $5,000, while the fourth case resulted in a plaintiff's verdict well below the amount offered by the defendant at mediation. Ms. Fanning speaks at legal conferences on defending lead-exposure claims and mold-exposure claims.

Ms. Fanning has earned the highest "AV" rating from Martindale Hubbell.

David M. Governo has a national reputation as a leader in the fields of indoor air quality and complex civil litigation. He counsels individuals and companies in business planning, such as risk management and regulatory compliance. Mr. Governo represents clients throughout the United States in litigation involving mold, lead, asbestos and chemical exposure claims.

In 2001, he developed and chaired Mealey's National Mold Litigation Conferences in California and Florida and he has been invited to present papers on mold at legal and trade conferences throughout the country. Mr. Governo has published articles on many aspects of mold including remediation, liability, insurance coverage and claims handling.

In 2004, Mr. Governo participated in several IAQ and Mold Conferences including: (1) the National Technical Workshops (sponsored by the New York City Department of Health and others) to develop guidelines to protect mold remediation workers, (2) "Mold, Liability and HVAC Engineering" at ASHRAE's Annual Meeting in Nashville, Tennessee and (3) the Harvard School of Public Health's Public Health Symposium on Mold on "How Mold Remediation is Challenging the Legal System."

Mr. Governo is a member of numerous bar and technical associations, including the American Society of Heating, Refrigerating and Air Conditioning Engineers (ASHRAE), the Indoor Air Quality Association, The American Indoor Air Quality Council, ACGIH, ASTM and the Editorial Board of *Indoor Environmental Connections*. He co-authored a text book titled *Toxic Torts: Law and Science Manual* (1996) and the *Toxic Tort and Medical Internet Directory* published by the Defense Research Institute.

Leslie O'Neal-Coble is a shareholder and a member of Greenberg Traurig's national construction law practice group. She has extensive experience handling indoor air quality issues including toxic mold and "sick building" claims. She was the lead attorney for the trial team in the Martin County Courthouse case, which led to a $24 million recovery for the County for design and construction defects that led to water intrusion, high humidity and mold growth. She has handled similar cases in Oregon and throughout Florida. She frequently counsels clients on methods to avoid water intrusion and mold problems and the remediation of mold in buildings. Ms. O'Neal-Coble has published numerous articles on "sick buildings" and mold litigation for national groups including the American Bar Association and the American College of Construction Lawyers. She has lectured on these topics to industry groups including the South Florida Homebuilders Association and the International Recovery Contractors Association. She is a member of and the former chair of the ABA Forum on the Construction Industry. She is a fellow in the American College of

Construction Lawyers, the American Bar Association Foundation and The Florida Bar Association. Ms. O'Neal-Coble earned her B.A. degree from the University of Florida and her J.D. degree from the University of Florida College of Law.

Stephen Pate, Esquire has been a partner since 1994 in the Houston office of Fulbright & Jaworski L.L.P. He focuses his litigation practice on construction litigation, first party insurance litigation, commercial coverage litigation, fraud, extracontractual and insurance litigation. Before joining the firm, Mr. Pate served as law clerk to Judge Joe J. Fisher, U.S. District Court. Mr. Pate is a past chair of the Property Insurance Committee of the Tort and Insurance Practice Section of the American Bar Association. He is a member of the Federation of Defense and Corporate Counsel and is chair of that body's Property Insurance Section. Pate was the Mid-Year Program Coordinator for the Tort and Insurance Section of the American Bar Association. He has served as a panelist at the Property Loss Research Bureau for the past several years. Pate has spoken on property insurance and bad faith issues at seminars presented by American Bar Association National Institutes, Loss Executives Association, the Defense Research Institute, the Federation of Insurance and Corporate Counsel, and the Mealy's Company. Pate serves on the Editorial Board of the *Bad Faith Law Reporter* and *The Insurance Coverage Litigation Reporter*. Pate authored the Chapter on "Inland Marine Insurance" for West's Insurance Coverage Litigation Series, and has authored the Property Insurance Survey for the *Tort and Insurance Practice Journal*. Mr. Pate is Co-editor of *Property Insurance Annotations: Fire and Extended Coverages* (3rd edition) (ABA 1998). Mr. Pate is a fellow of the Houston and the Texas Bar Foundations and has served as Director of both the Texas and Houston Young Lawyers Associations. He is listed in Who's Who in America, Who's Who in America Law, Who's Who in the South and Southwest, and Who's Who in the World. While in law school, he was research editor for the Vanderbilt Journal. Mr. Pate received a B.A., magna cum laude, in 1980 and a J.D. in 1983 from Vanderbilt University. He is a member of Phi Beta Kappa. He has tried over forty cases in twelve different Texas counties.

Patrick J. Perrone, Esquire is a partner at McCarter, LLP in Newark, New Jersey. His practice is concentrated in civil litigation, with particular emphasis on the defense of mold claims involving builders, property owners, and building product manufacturers. Mr. Perrone is currently defending a number of mold claims and has written and spoken extensively about mold litigation. Mr. Perrone has appeared in numerous courts throughout the country and has handled over 100 building products cases. In addition to his trial experience, Mr. Perrone has performed extensive appellate work, including appeals in the New York Court of Appeals, and numerous Federal Courts of Appeal. Mr. Perrone is a member of the American Bar Association, the New Jersey State Bar Association, the New Jersey Builders Association, and the National Association of Home Builders. Mr. Perrone serves on the Legal Action Committee of the New Jersey Builders Association and is a member of the Mealey Publications 2003 Toxic Tort Advisory Council.

Douglas F. Schleicher, Esquire is a partner at the Philadelphia law firm of Klehr, Harrison, Harvey, Branzburg & Ellers and chairs the firm's Environmental Group. He is a leading practitioner in the areas of mold remediation and indoor air quality, as well as in Brownfields redevelopment, transactional environmental issues and environmental litigation. Recently, Mr. Schleicher has represented the owner of an apartment complex in both emergent and class action litigation arising from the condemnation of buildings due to the presence of mold, as well as in the remediation, sale and insurance coverage negotiations for the facility. In addition, he has worked with numerous building and apartment owners and managers on numerous matters arising from the actual or suspected presence of mold, including investigation and remediation, tenant interaction, and development of pro-active contract, due diligence and insurance strategies. He has written articles and participated as a faculty member at a number of seminars addressing mold. Mr. Schleicher is a member of the American Bar Association's Section of Environment, Energy and Resources; the Pennsylvania Bar Association's Section on Environmental, Mineral & Natural Resources; the Philadelphia Bar Association's Environmental Law Section; the Building Industry Association; and the Apartment Association of Philadelphia. He received his J.D. from the University of Michigan Law School and his A.B. from Princeton University.

Cynthia J. Stephens has defended complex toxic torts cases, including mold, lead poisoning, asbestos, chemical exposure, and indoor air quality claims, for the past thirteen years. She and her partner tried one of the first Massachusetts lead poisoning cases and obtained a defense verdict. She has also participated in the successful mediation of numerous cases. She has spoken on the topic of toxic torts across the county at large national conferences, business organizations, and county groups. Ms. Stephens is also the author of several articles, papers on the EPA's lead regulations, and legal trends in mold and lead cases. She received a JD cum laude from Suffolk University School of Law, a BA cum laude from the University of Massachusetts in Amherst, and is licensed to practice law in Massachusetts and New Hampshire.

Richard J. Valladares, Esquire is an associate at the law firm of McCarter & English, LLP.

Mr. Valladares' practice consists primarily of complex commercial litigation including insurance coverage, building products litigation, class actions, commercial contracts, product liability and toxic torts.

Bryan D. Weiss, Esquire is an associate and member of the Environmental Group of the Philadelphia law firm of Klehr, Harrison, Harvey, Branzburg & Ellers. He concentrates his practice on environmental aspects of real estate transactions, corporate due diligence and regulatory matters. He also dedicates a significant portion of his time to environmental litigation, including mold and other toxic tort claims. Mr. Weiss' employment background includes working for a large multi-national environmental engineering firm. He also practiced as an in-house attorney for a specialty liability insurance carrier. In that capacity, Mr. Weiss dealt with a variety of professional liability insurance programs related to building design and construction projects. He also developed significant experience dealing with complex insurance coverage issues involving environmental contamination. Mr. Weiss is a member of the American Bar Association's Tort and Insurance Practice Section; the Pennsylvania Bar

Association's Section on Environmental, Mineral & Natural Resources Law; and the Philadelphia Bar Association's Environmental Law and Insurance Programs committees. He received his J.D. and Master of Studies in Environmental Law from the Vermont Law School.

Dennis I. Wilenchik, Esquire is the founding shareholder of the Phoenix, Arizona law firm of Wilenchik & Bartness P.C. He has practiced extensively for over twenty-five years, principally in the areas of real estate, construction and general commercial litigation. He is a nationally certified Civil Trial Advocate with the National Board of Trial Advocacy and a former chairman of the Trial Practice Section of the State Bar of Arizona, President of the Maricopa County Bar Association Young Lawyers Division, Secretary of the State Bar of Arizona Civil Practice and Procedure Committee, and member of the Superior Court's Civil Study committee, and editorial board member of the *Arizona Bar Journal*. He is a former partner in the international law firm of Squire, Sanders & Dempsey, and a deputy County Attorney and Judge Pro Tem of the Superior Court of Maricopa County, and is a frequent lecturer on trial tactics, evidence, mold and construction defect issues.

Index